Conference Proceedings of the Society for Experimental Mechanics Series

The Conference Proceedings of the Society for Experimental Mechanics Series presents early findings and case studies from a wide range of fundamental and applied work across the broad range of fields that comprise Experimental Mechanics. Series volumes follow the principle tracks or focus topics featured in each of the Society's two annual conferences: IMAC, A Conference and Exposition on Structural Dynamics, and the Society's Annual Conference & Exposition and will address critical areas of interest to researchers and design engineers working in all areas of Structural Dynamics, Solid Mechanics and Materials Research.

Frank Gardea • Kunal Mishra • Michael Keller
Editors

Mechanics of Composite, Hybrid and Multifunctional Materials, Volume 5

Proceedings of the 2023 Annual Conference & Exposition on Experimental and Applied Mechanics

Springer

Editors
Frank Gardea
U.S. Army Research Laboratory
College Station, TX, USA

Kunal Mishra
Corning Incorporated
Painted Post, NY, USA

Michael Keller
University of Tulsa
Tulsa, OK, USA

ISSN 2191-5644 ISSN 2191-5652 (electronic)
Conference Proceedings of the Society for Experimental Mechanics Series
ISBN 978-3-031-50477-8 ISBN 978-3-031-50478-5 (eBook)
https://doi.org/10.1007/978-3-031-50478-5

This Springer imprint is published by the registered company Springer Nature Switzerland AG
The registered company address is: Gewerbestrasse 11, 6330 Cham, Switzerland

Preface

Mechanics of Composite, Hybrid and Multifunctional Materials represents one of five volumes of technical papers presented at the 2023 SEM Annual Conference & Exposition on Experimental and Applied Mechanics organized by the Society for Experimental Mechanics scheduled held on June 5–8, 2023. The complete proceedings also include volumes on: Additive and Advanced Manufacturing, Advancement of Optical Methods in Experimental Mechanics, Dynamic Behavior of Materials, Fracture and Fatigue, Inverse Problem Methodologies, Machine Learning and Data Science, Mechanics of Biological Systems and Materials, Mechanics of Composite and Multifunctional Materials, Residual Stress, Thermomechanics and Infrared Imaging, and Time-Dependent Materials.

The commercial market for composite continues to expand with a wide range of applications from sporting equipment to aerospace vehicles. This growth has been fueled by new material developments, greater understanding of material behaviors, novel design solutions, and improved manufacturing techniques. The broad range of applications and the associated technical challenges require an increasingly multidisciplinary and collaborative approach between the mechanical, chemical, and physical sciences to sustain and enhance the positive impact of composites on the commercial and military sectors.

New materials are being developed from recycled source materials, leading to composites with unique properties and more sustainable sources. Existing materials are also being used in new and critical applications, which requires a deeper understanding of material behaviors and failure mechanisms on multiple length and time scales. In addition, the unique properties of composites present many challenges in manufacturing and in joining with other materials. New testing methods must be developed to characterize the novel composite properties, to evaluate application and product life cycle performance, as well as to evaluate impacts and merits of new manufacturing methods.

This volume presents early research findings from experimental and computational investigations related to the processing, characterization, and testing of composite, hybrid, and multifunctional materials.

College Station, MD, USA — Frank Gardea
Painted Post, NY, USA — Kunal Mishra
Tulsa, OK, USA — Michael Keller

Contents

Deformation and Flow Characterization of Hybrid Woven Composite Material

HyeokJae Lee and Ashraf Bastawros

Abstract Woven and textile composite materials exhibit a highly inhomogeneous deformation field with the complex interaction between the deformation and failure mechanisms, which controls the structure-level specific absorption energy. This work attempts to provide insight into the deformation, flow, and post-peak characteristics of hybrid weave composite materials by examining the deformation characteristics of off-axis loading of a woven laminate. Full-field displacement measurements by digital image correlations (DIC) and global coordinate axial strains by non-contact video extensometer were employed. We proposed to utilize the measured incremental plastic strain field over the reduced Moher strain plane to highlight the material non-associative deformation, flow characteristics, and the kinematics of shear band formations and their relation to the macroscopic homogenized flow response of the material. We noticed that the *weave periodicity* imposes an additional microstructure length scale that requires 4–6 periods to accommodate a localized failure band. The macroscopically measured mechanical properties and the observed cell-level deformation mechanisms will support the development of constitutive input parameters for the deformation and flow characterization of user-material subroutines.

Keywords Woven composites · Shear band · Digital image correlation · Video extensometer · Reduced yield surface

Introduction

Plain weave composites have gained significant attention in various applications, such as military armor, boat/yacht hulls, automotive parts, bridge components, wind turbines, thermal systems, and aircraft parts, thanks to their numerous advantages. These advantages include high mechanical in-plane properties, fracture toughness, transverse strength, damage tolerance, controllable delamination suppression, and impact resistance [1, 2]. Plain weave composites, including Carbon/Kevlar, E-glass/Polyester, and E-glass/Epoxy composites, have been widely used in practical applications [3, 4]. The Carbon/kevlar® plain weave composite is particularly noteworthy and has been employed by NASA in lightweight Vertical Take-off and Landing (eVTOL) vehicles with electric propulsion and rotorcraft as an impact-resistant material, utilizing the high strength of carbon and the high toughness of Kevlar® [5, 6]. However, a comprehensive understanding of the mechanical characteristics and deformation mechanisms of this composite is essential to fully exploit its properties.

Plain weave composites have been extensively analyzed both analytically and experimentally by various researchers to understand their mechanical properties. De Carvalho et al. [7] developed a numerical model to predict the stiffness and failure mode of woven composites, while Green et al. [8] presented a 3D woven composite model that accounts for realistic unit cell geometry to prevent overestimation of stiffness and strength. Han et al. [9] addressed the computationally expensive issue of nonlinear woven Representative Volume Element (RVE) of multiscale simulation of woven composites structure using Self-consistent Clustering Analysis (SCA). Baker et al. [10] utilized Finite Element (FE) simulation and evolutionary algorithm to optimize elastic properties and weaving patterns of woven composites and to examine the impact of various geometry factors. Quinn et al. [11] studied the failure and strain distribution of 3D woven composites under uniaxial tension, while Yu et al. [12] investigated the failure mechanisms of 3D woven composites under both static and cyclic fatigue uniaxial loading. Atas et al. [13] explored the impact of yarn angle on energy absorption under impact. These studies have been valuable in estimating and understanding the mechanical properties of woven composites. However, more in-depth experimental studies are needed under off-axis loading due to the complexity of the composite structure. This is essential as plain weave composites are widely used in many shell structures and are often subject to off-axis loading [14].

H. Lee · A. Bastawros (✉)
Department of Aerospace Engineering, Iowa State University, Ames, IA, USA
e-mail: hyeokjae@iastate.edu; bastaw@iastate.edu

F. Gardea et al. (eds.), *Mechanics of Composite, Hybrid and Multifunctional Materials, Volume 5*,
Conference Proceedings of the Society for Experimental Mechanics Series, https://doi.org/10.1007/978-3-031-50478-5_1

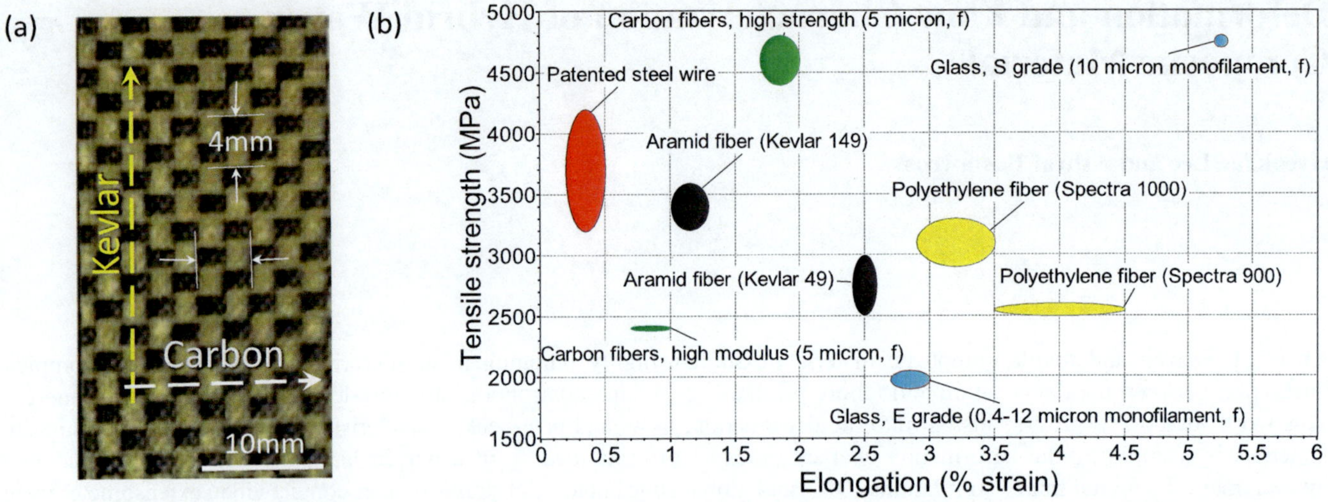

Fig. 1 (**a**) The microstructure of carbon/Kevlar® plain weave composite, where horizontal dark fibers are carbon, and vertical bright fibers are Kevlar®. A single weave period comprises a 4 mm bundle of both carbon and Kevlar® fibers. (**b**) An Ashby-Plot showing the relative strength of the carbon fiber and the toughness of the Kevlar® fiber [17]

The primary objective of this study was to investigate the behavior of Carbon/Kevlar® plain weave composite under 45° off-axis uniaxial tensile loading. The study aimed to provide insight into the deformation and flow characteristics, as well as shear band nucleation, formation, and propagation. To achieve this, global coordinate axial and transverse strain measurements were taken using a non-contact video extensometer. Furthermore, 1-D and 2-D displacement and strain maps of the specimen were obtained using digital image correlation (DIC) to reveal the non-associative deformation, flow characteristics, and kinematics of the shear band. This study employs the reduced yield surface initially proposed by Rice et al. [15] and subsequently applied by Bastawros and Kim [16] for orthotropic materials to highlight the non-associative deformation of the specimen. The results of this study not only provide insights into plain weave composites but also suggest ways to improve their performance.

Experimental Background (Fig. 1)

The tested material is a hybrid carbon-Kevlar® plain-weave fabric, which consisted of 3k-sized carbon fibers in the warp direction and 3k-sized Kevlar® fibers in the fill direction. An 8-ply laminate is fabricated with 105/205 West system epoxy matrix. The direction of the carbon fibers oriented in the 1-direction, Kevlar® fibers in the 2-direction, and thickness in the 3-direction, resulting in an orthotropic material. Dog-bone-shaped specimens were fabricated in accordance with ASTM standards for conducting 45° off-axis tensile tests [18]. The global coordinate strains were measured using a non-contact video extensometer, while full-field displacement and strains were obtained using 2D-DIC [19]. In this study, the sample is subjected to tensile testing using a universal screw-driven testing frame (Instron 5969) under displacement control at a rate of 20 μm/s, while utilizing wedge-shaped frictional grips (Fig. 2).

Digital image correlation (DIC) is a non-contact and non-interferometric optical technique that allows for the measurement of full-field displacements and strains at consecutive stages of external loading [19–29]. The technique provides the full-field deformation over time, which is valuable in analyzing shear band mechanisms. The method involves comparing two digital images taken of a random speckle pattern on the surface of a test specimen before and after applying incremental deformation, using the concept of cross-correlation to estimate the shifts in datasets. A unique speckle pattern is generated on the specimen surface by applying paint spray, and a group of neighboring pixels or subsets is used to track the motion of points within a region of interest. The displacement components are calculated between two corresponding windows with the maximum correlation. Selecting the appropriate window size for a region of interest is critical to obtain high-quality results. An improper window size may produce over-averaged data or noisy displacement data. To show the micro-behavior of the specimen, a window size smaller than one weave period of the specimen is necessary. Basler acA4096-11 camera was used to capture the testing images over time for DIC analysis. The camera frame rate was 12 fps, the resolution was 4096 px × 2160 px, the pixel size was 3.45 μm × 3.45 μm, the recording rate was 0.5 fps, which corresponds to a macrostrain of about 0.04%. In this study,

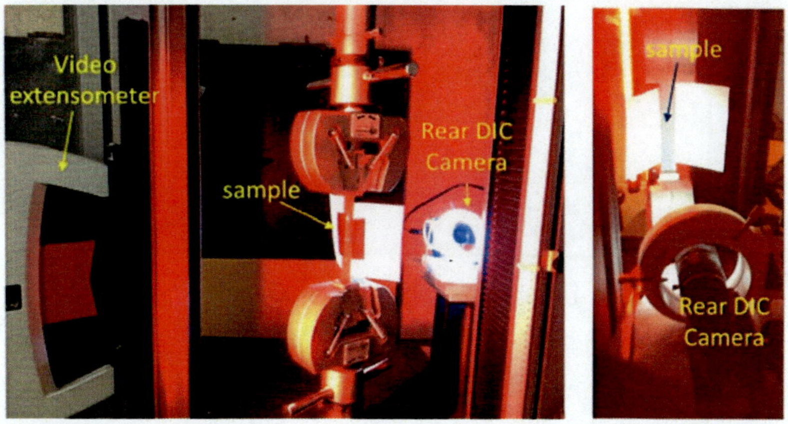

Fig. 2 The experimental arrangement for the off-axis tensile test is illustrated, demonstrating the implementation of a video extensometer on the front surface of the specimen, in conjunction with the capture of digital image correlation (DIC) images on the rear surface of the specimen

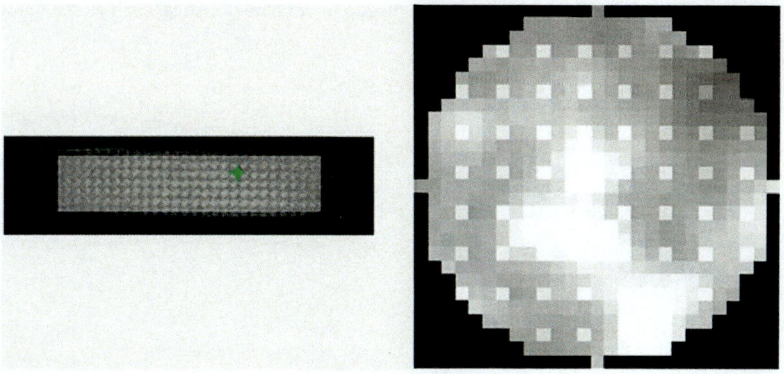

Fig. 3 Digital image correlation analysis. Left: the whole specimen gauge. Right: a DIC window of 25pixel with analysis seed of every other third pixels, providing 20 analysis points per weave period

an optical magnification of 40 μm/pixel was utilized with a correlation window of 25 pixels. This provides a correlation window of about 1 mm, which less than half the weave period of 2.4 mm. An analysis grid point is chosen every 3-pixels, providing about 20 analysis points per weave period (Fig. 3).

This study utilized a non-contact video extensometer to accurately measure global coordinate axial and transverse strains during the test, while ensuring the specimen was not affected. By detecting the motion of two dots on the specimen, changes in the axial and transverse displacement were precisely measured, providing undistorted data for the macroscopic axial and transverse strains, as the dots were outside the shear bands.

To analyze the non-associative deformation and flow characteristics of the material during deformation, a reduced yield criterion was employed. This criterion is suitable for anisotropic, incompressible, homogeneous, and rigid/plastic materials. The Mohr plane can be expressed in terms of the strain variables $(\varepsilon_x - \varepsilon_y)/2$ and ε_{xy} as shown in Fig. 4, whereby each point on the Mohr stain plane corresponds to the unique strain components at a material point on the sample surface. Upon reaching yield, the strain increment at each material point may be in the direction normal to the yield surface (assuming associated flow, and proportional loading) and thereby the corresponding stress state could be identified. By tracking the progression of these trajectories of the strain increments, it is possible to discern the hardening nature (i.e., isotropic vs. orthotropic hardening), the influence of orthotropic behavior on the flow characteristics, as well as the potential evolution of the plastic Poisson's ratio with plastic strain.

Analysis

The measured axial load was converted to nominal tensile stress, σ_x. The video extensometer provided measurements for both the axial strains and the transverse strains. The in-plane maximum shear stress, $\tau_{\max}$ was calculated by using

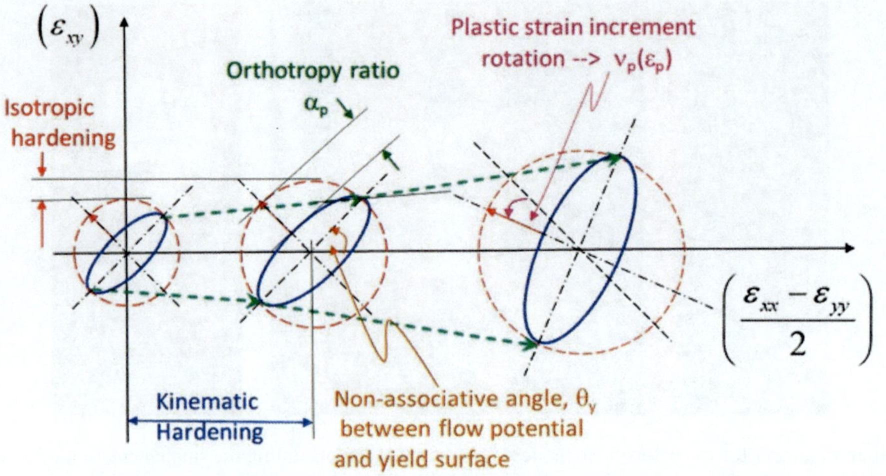

Fig. 4 The application of a reduced Mohr strain representation was employed to highlight the orthotropic flow behavior of woven composites, and it is accompanied by visual illustrations, which aid in identifying various plastic flow characteristics, such as the hardening nature, orthotropic ratio, and variation of plastic Poisson's ratio with plastic strain

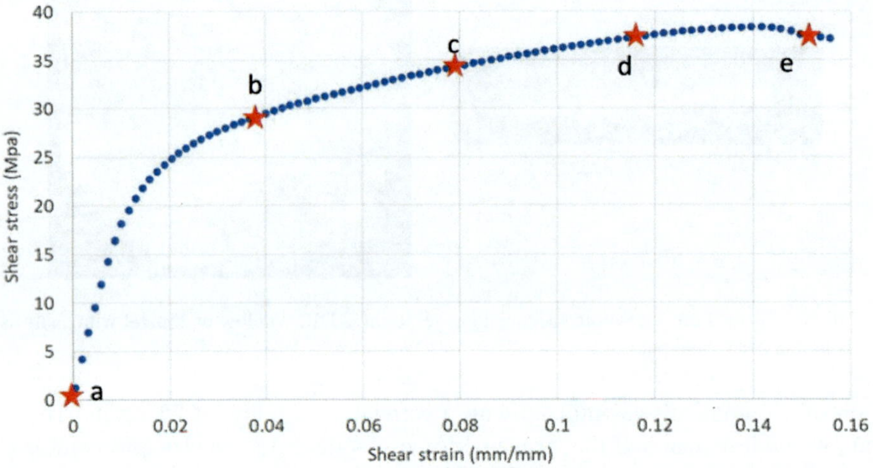

Fig. 5 Maximum in-plane shear stress-strain curve for the 45° off-axis tensile test. The a, b, c, d, and e points correspond to (a), (b), (c), (d), and (e) of Fig. 6

$$\tau_{\max} = (\sigma_x - \sigma_y)/2 = \sigma_x/2 \tag{2}$$

Here, σ_x is the loading direction stress, and σ_y is the transverse loading direction stress, where $\sigma_y = 0$ for uniaxial loading. The maximum shear strain was calculated using

$$\frac{\gamma_{\max}}{2} = \sqrt{\left(\frac{\varepsilon_x - \varepsilon_y}{2}\right)^2 + \left(\varepsilon_{xy}\right)^2} \tag{3}$$

$\gamma_{\max}$ is the maximum in-plane shear strain, ε_x is the loading direction strain, ε_y is the transverse loading direction strain, and γ_{xy} is the in-plane shear strain, which is negligibly small. So, the maximum in-plane shear strain was approximately calculated using

$$\gamma_{\max} = \varepsilon_x - \varepsilon_y \tag{4}$$

The shear stress-strain graph is shown in Fig. 5. The shear stress was increased linearly in the elastic region. However, in the plastic region, it showed an initial nonlinear increased, followed by extended linear hardening until it reached the specimen's ultimate shear strength with plateau stress. During the 45° off-axis tensile test, an extended range of hardening

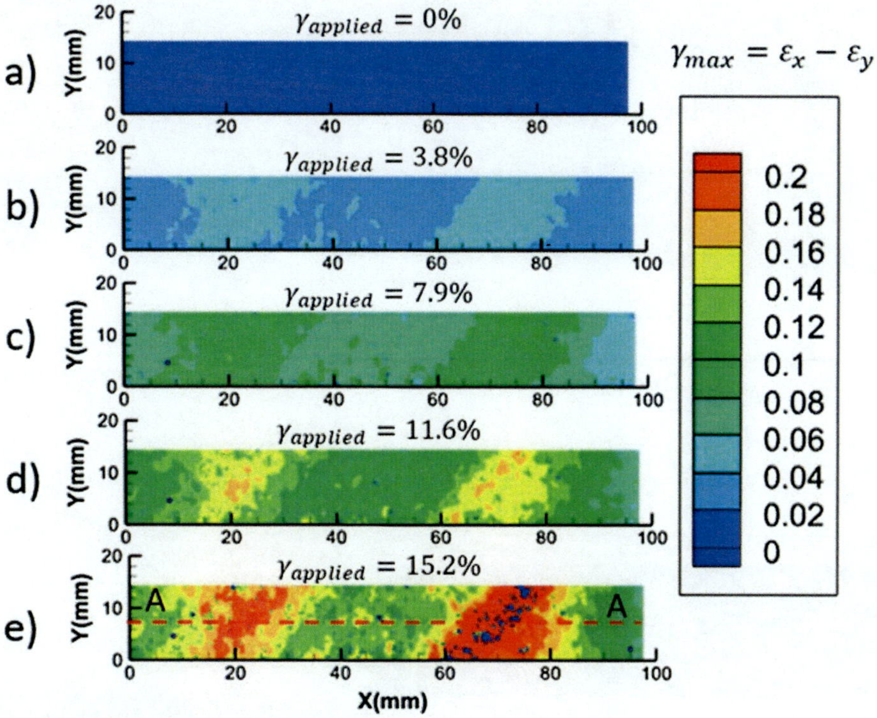

Fig. 6 2D DIC maps for the maximum in plane shear strain component at different loading levels, marked on Fig. 5 by the letters **a**–**e**. This range cover the transition from linear to nonlinear regime. Two shear bands were formed under the off-axis loading. The shear bands were initiated before reaching the ultimate strength. The global shear strain at each steps are also noted

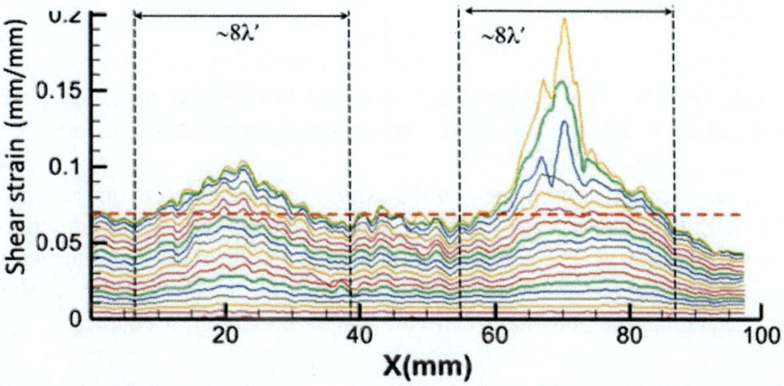

Fig. 7 Maximum in plane shear strain evolution across line A-A, marked on Fig. 6. The width of the shear bands is about eight times the weave period, or extending to about 5–6 weave periods on the rotated fabric axes

was observed until approximately 14% shear strain, after which the material underwent softening and exhibited post-peak stress behavior. Those behaviors are closely related to the underlying localized shear band deformation mechanism.

Figure 6 illustrates the 2-D DIC maximum in-plane shear-strain map ($\gamma_{\max} = \varepsilon_x - \varepsilon_y$) where each subfigure (a) through (e) corresponds to the points a, b, c, d, and e in Fig. 5. The map clearly indicates the formation and development of two distinct shear bands within the specimen.

In Fig. 7, the growth of shear strain along the dotted line A-A in Fig. 6 is presented. The shear strains continued to increase until they reached a saturation point of about 7%. Upon reaching a critical level of shear strain, which is postulated to be associated with the shear strain limit of the matrix, the formation of localized shear bands was observed. However, outside the region of the localized shear bands, the strain field appeared to remain nearly constant with no additional significant increase. The width of these localized shear deformation was approximately eight times the weave period, which is equivalent to 5–6 weave periods on the 45° rotated plane, parallel to the fabric coordinate. A single weave period is 4 mm in length and consists of one bundle of carbon fibers and one bundle of Kevlar® fibers.

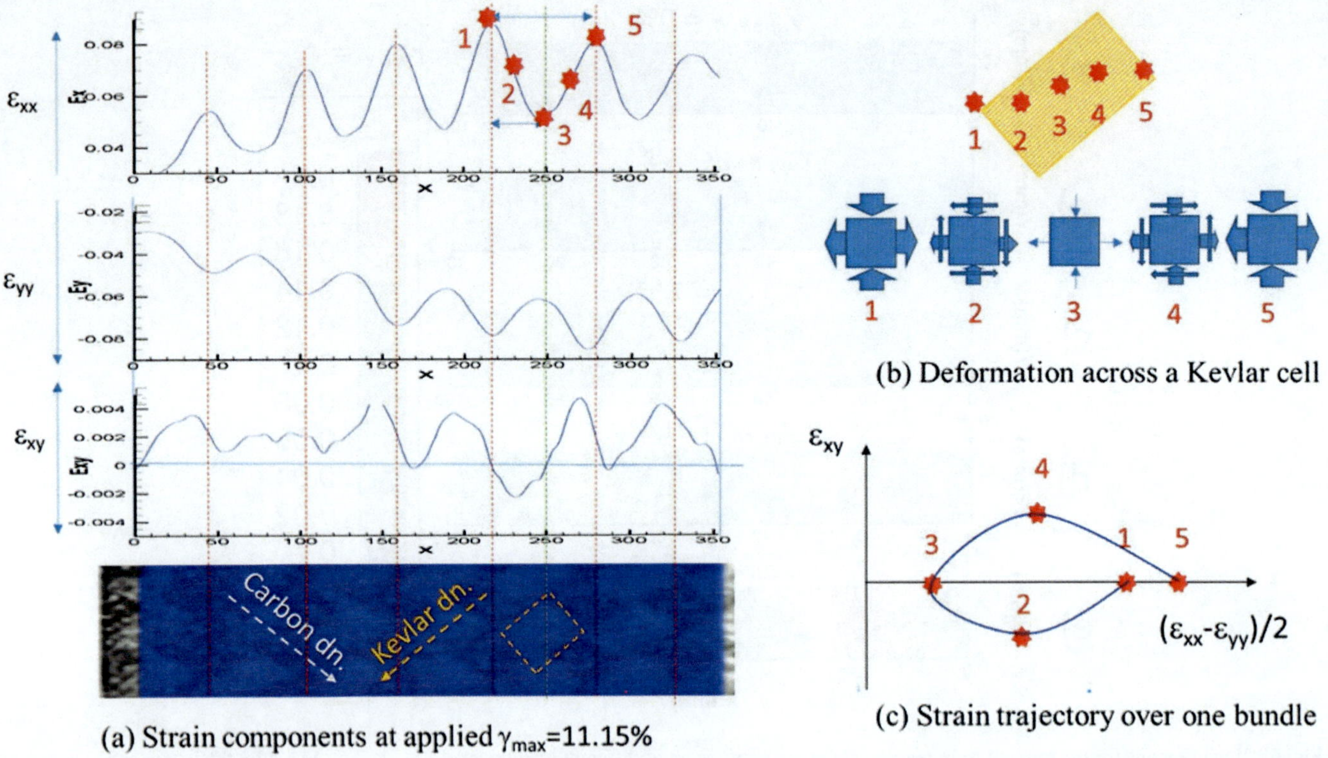

Fig. 8 The deformation of a single bundle at a macroscopic maximum shear strain of 11.15%, near the plateau stress level. (**a**) The distribution of the in-plane strain components along a parallel to the loading axis, which exhibits oscillations coinciding with the crimping points of the carbon and Kevlar® tows. (**b**) The assignment of peak locations and corresponding stress states within a Kevlar® cell. (**c**) A reduced Mohr strain representation depicting the full trajectory from material points 1 to 5

Figure 8 shows the cell-level deformation represented on reduced Mohr strain plane at a macroscopic applied strain of $\gamma_{max} = 11.15\%$. The representation is selected near the start of the saturation stress on the macroscopic stress-strain curve of Fig. 5.

Figure 8a illustrates the in-plane strain components, which exhibit periodic fluctuations around the crimping points of the carbon-Kevlar® weave. The marked points 1–5 on a Kevlar® weave reveal principal stretch along the loading direction and contraction along the transverse direction at crimp points and within the middle of the cell. Furthermore, the traversal from one material point of a principal state of stretch to another produces a state of maximum in-plane shear. Tracking the complete strain trajectory within a single cell has the potential to provide insight into the flow characteristics of the composite material throughout the loading process.

Figure 9 summarizes the evolution of the strain trajectory across a single weave period within the localized shear band of Fig. 6e. Figure 9a shows the macroscopic shear stress-shear strain curve. The time steps between analyzing point 1 and point 9 are 50 s, between points 9 and 10 is 20 s, from 10 to 13 is 10 s, and from 13 to 14 is 40 s. Figure 9b shows the evolution of the strain trajectory over the a single weave period, within the shear band of Fig . 6e. The strain trajectory grows in a self-similar way but with shifting center until step 9. During this linear hardening phase, the average strain trajectory shows a self similar growth with isotropic hardening can be ascertained from such behavior. Reaching the peak stress during loading steps 9–11, a subtle rotation of the strain trajectory axes could be observed, which indicate changes of the non-associative angle between the flow potential and the yield surface. This would lead to variation of the plastic Poisson's ratio at this level of local strain within the shear band. At loading steps 11–14, the strain trajectory is completely changing in an inconsistent manner, coinciding with the formation of a macroscopically observable localized shear band and attaining the peak stress level.

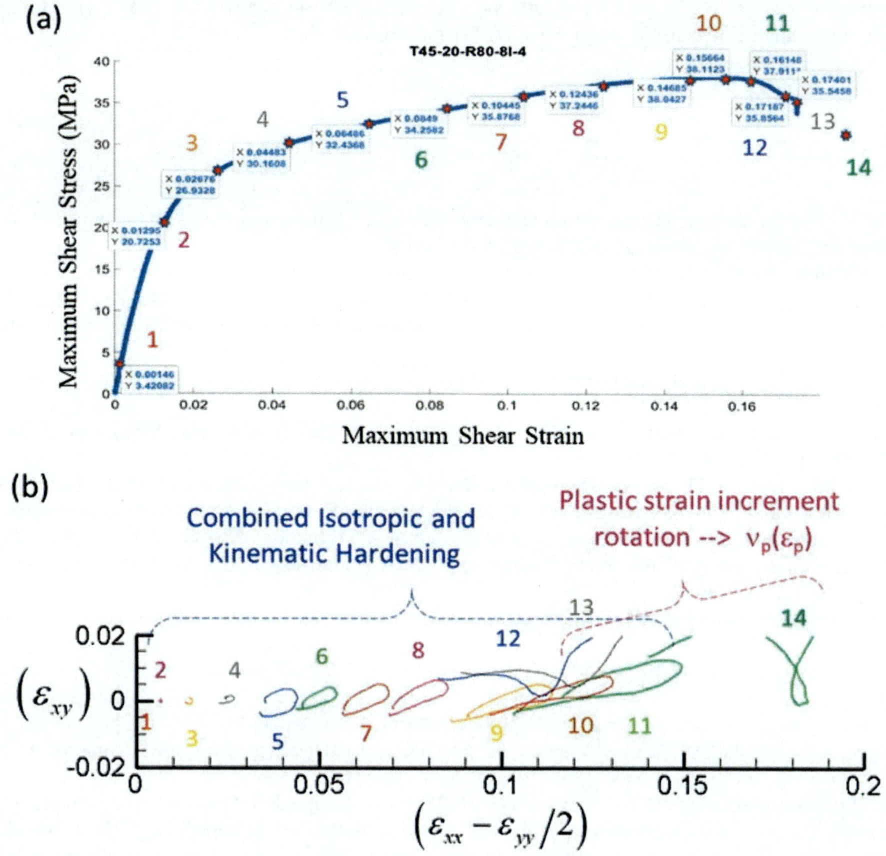

Fig. 9 The deformation history within a single weave period within the shear band of Fig. 6e, depicting the evolution of the strain increment within the shear band during the hardening phase of deformation until reaching the peak stress. (**a**) The macroscopic shear stress-shear strain curve, showing the analysis points. (**b**) Experimentally measured strain components on reduced Mohr strain plane across a single weave period within the localized shear band of Fig. 6e, highlighting the evolution of the flow potential and its spatial rotation near the incipient formation of shear band (steps 9–11)

Conclusion

The macroscopic shear stress-shear strain curve is derived from the 45° off-axis tensile loading for a carbon-Kevlar® hybrid woven laminate. This curve exhibits an extended linear hardening regime up to 14% of macroscopic shear strain, before the deformation becomes localized in the form of a shear band. Digital image correlation measurements taken during loading reveal several important characteristics of the plastic flow and deformation behavior of this woven composite. In particular, the in-plane strain components exhibit fluctuations that correspond to the periodicity of the carbon-Kevlar® crimping points. The strain states within one weave periodicity traverse states of principal stretches and maximum shear strain. Tracking the evolution of the strain trajectory on reduced Mohr strain space shows that it grows in a self-similar manner with a progressively increasing locus. These experimental observations indicate that the macroscopically observed linear hardening regime is associated with isotropic hardening till reaching the peak stress. The formation of the localized shear band is associated with a rotation of the strain trajectory, indicating an evolution of the non-associative angle between the flow potential and yield surface. Beyond the peak stress, the progressive macroscopic softening resulted in a non-correlated strain trajectory. By normalizing the Mohr strain space appropriately, it is possible to assess the orthotropic ratio and many of the parameters employed in the assumed yield function and plastic flow potential. This approach enables a deeper understanding of the deformation behavior of these hybrid orthotropic materials and provides quantification for many of the flow potential assumed parameters. Such a framework will enable the design and calibration of many new concepts for deformation and energy absorption structures.

Acknowledgments This work was supported by NASA grant numbers 80NSSC21M0143 and 80NSSC22M0241, with Dr. Robert K. Goldberg as the technical officer, and by Iowa State University through the T.W. Wilson Professorship.

References

1. Bogdanovich, A.E. (ed.): Advancements in manufacturing and applications of 3D woven preforms and composites. In: Proceeding of the 16th International Conference on Composites Materials (ICCM-16), Kyoto (2007)
2. Lomov, S.V., et al.: A comparative study of tensile properties of non-crimp 3D orthogonal weave and multi-layer plain weave E-glass composites. Part 1: materials, methods and principal results. Compos. A: Appl. Sci. Manuf. **40**(8), 1134–1143 (2009)
3. Pandya, K.S., Veerraju, C., Naik, N.K.: Hybrid composites made of carbon and glass woven fabrics under quasi-static loading. Mater. Des. **32**(7), 4094–4099 (2011)
4. Faizal, M.A., Beng, Y.K., Dalimin, M.N.: Tensile property of hand lay-up plain-weave woven e-glass/polyester composite: curing pressure and ply arrangement effect. Borneo Sci. **19**, 27–34 (2006)
5. Littell, Justin, Jacob Putnam, and Robin Hardy. The Evaluation of Composite Energy Absorbers for Use in UAM eVTOL Vehicle Impact Attenuation. No. NF1676L-31347. 2019
6. Jackson, K.E., Fasanella, E.L., Littell, J.D.: Development of a continuum damage mechanics material model of a graphite-Kevlar (registered trademark) hybrid fabric for simulating the impact response of energy absorbing Kevlar (registered trademark) hybrid fabric for simulating the impact response of energy absorbing. In: AHS International Annual Forum & Technology Display. No. NF1676L-26695 (2017)
7. De Carvalho, N.V., Pinho, S.T., Robinson, P.: Numerical modelling of woven composites: biaxial loading. Compos. A: Appl. Sci. Manuf. **43**(8), 1326–1337 (2012)
8. Green, S.D., et al.: Mechanical modelling of 3D woven composites considering realistic unit cell geometry. Compos. Struct. **118**, 284–293 (2014)
9. Han, X., et al.: Efficient multiscale modeling for woven composites based on self-consistent clustering analysis. Comput. Methods Appl. Mech. Eng. **364**, 112929 (2020)
10. Bakar, I.A.A., et al.: Optimization of elastic properties and weaving patterns of woven composites. Compos. Struct. **100**, 575–591 (2013)
11. Quinn, J.P., McIlhagger, A.T., McIlhagger, R.: Examination of the failure of 3D woven composites. Compos. A: Appl. Sci. Manuf. **39**(2), 273–283 (2008)
12. Yu, B., et al.: 2D and 3D imaging of fatigue failure mechanisms of 3D woven composites. Compos. A: Appl. Sci. Manuf. **77**, 37–49 (2015)
13. Atas, C., Liu, D.: Impact response of woven composites with small weaving angles. Int. J. Impact Eng. **35**(2), 80–97 (2008)
14. Tang, J., et al.: Failure analysis of plain woven glass/epoxy laminates: comparison of off-axis and biaxial tension loadings. Polym. Test. **60**, 307–320 (2017)
15. Rice, J.R.: Plane strain slip line theory for anisotropic rigid/plastic materials. J. Mech. Phys. Solids. **21**(2), 63–74 (1973)
16. Bastawros, A.F., Kim, K.S.: Experimental analysis of near-crack-tip plastic flow and deformation characteristics (I): polycrystalline aluminum. J. Mech. Phys. Solids. **48**, 67–98 (2000)
17. Zoghi, M. (ed.): The International Handbook of FRP Composites in Civil Engineering. CRC Press, Boca Raton (2013)
18. ASTM International: Standard Test Method for Tensile Properties of Plastics. ASTM International, West Conshohocken (2014)
19. Blaber, J., et al.: Ncorr: open-source 2D digital image correlation Matlab software. Exp. Mech. **55**, 1105–1122 (2015)
20. Chu, T.C., et al.: Applications of digital image correlation techniques to experimental mechanics. Exp. Mech. **25**(3), 232–244 (1985)
21. Bart-Smith, H., et al.: Compressive deformation and yielding mechanisms in cellular Al alloys determined using X-ray tomography and surface strain mapping. Acta Mater. **46**(10), 3583–3592 (1998)
22. Bastawros, A.F., McManuis, R.: Case study: use of digital image analysis software to measure non-uniform deformation in cellular aluminum alloys. Exp. Tech. **22**, 35–37 (1998)
23. Bastawros, A.F., et al.: Experimental analysis of deformation mechanisms in a closed-cell aluminum alloy foam. J. Mech. Phys. Solids. **48**, 301–322 (2000)
24. Bastawros, A.F., Evans, A.G.: Deformation heterogeneity in cellular Al alloys. Adv. Eng. Mater. **2**(4), 210–214 (2000)
25. Antoniou, A., Bastawros, A.F.: Deformation characteristics of tin-based solder joints. J. Mater. Res. **18**(10), 2304–2309 (2003)
26. Antoniou, A., et al.: Experimental analysis of compressive notch strengthening in closed-cell aluminum alloy foam. Acta Mater. **52**(8), 2377–2386 (2004)
27. Antoniou, A., et al.: Deformation behavior of a zirconium based metallic glass during cylindrical indentation: in-situ observations. Mater. Sci. Eng. **394**(1–2), 96–102 (2005)
28. Antoniou, A., et al.: Experimental observations of deformation behavior of bulk metallic glasses during wedge-like cylindrical indentation. J. Mater. Res. **22**(2), 514–524 (2007)
29. Hwang, H., et al.: Application of digital image correlation for multiscale biomechanics. In: Neu, C., Genin, G. (eds.) Handbook of Imaging in Biological Mechanics, pp. 141–150. CRC Press, Boca Raton (2014)

Development of Recycled Alumix431-Based Composites Reinforced with TiB_2, TiC, and B_4C Fine Ceramic Powders for Aircraft Applications Produced by a Combined Method: Sintering + Forging

Eduardo José Bernardes, Fabio Gatamorta, Isabella Carvalho Lancini, Ibrahim Miskioglu, and Emin Bayraktar

Abstract In this chapter, the microstructural formation and static/cyclic compression behavior of recycled Alumix (aluminum alloy) matrix hybrid composites reinforced with TiB_2, TiC, and B_4C are studied. It is aimed as an alternative to traditional alloys/composites used in the aeronautical industry. These composites are generally produced by using a combined sintering + forging process. The static and dynamic properties are evaluated in detail, taking into account the relevant scanning electron microscopy microstructures (including the distribution of reinforced elements).

Keywords Recycled and mixed aluminum · Hybrid metal composites · TiB · TiC · B_4C static-cyclic time depending tests · SEM analyses

Introduction

Innovative new design of high-strength, high-toughness hybrid metal matrix composites for use in the manufacturing of certain pieces for thermos-compressor in aircraft engineering. Some of the requirements for these pieces are remarkable mechanical properties (static, cyclic, time-dependent) and a healthy microstructure. In fact, these new composites [1–7] are also called multifunctional. The behaviour of these materials are affected by the heterogeneous structures. In this work, a new innovative hybrid composite has been produced, with different fine ceramic reinforcements [6–10]. Our research project that is going on has shown that the reinforcements of the fine TiB_2, TiC and B_4C (1–10 μm) ceramic particles, minor nano molybdenum and fine copper particles improve mechanical properties of these composites [5–12]. For this reason, a special combined method called "Sinter + Forging" was used for their production. Another method called "thixoforming/semisolid sintering" for some of the composites to compare certain properties such as toughness microstructure, etc. [8–14].

In the present work, the static and dynamic behaviours of these composites were evaluated to compare them with the former studies carried out by the same research group.

Experimental Conditions

As a matrix for these composites, "Alumix 431 + A1050" was chosen from the recycled chips supplied by a Brazilian aeronautic company. Recycled aluminium (AA 431) chips were gas atomized after mixing with pure A1050 (50 wt%) prepared by high-energy milling in a planetary ball mill under an inert argon atmosphere to prevent oxidation of the powders

E. J. Bernardes · I. C. Lancini
UNICAMP, University of Campinas, FEM, Materials Science, Campinas, SP, Brazil

F. Gatamorta
State University of Rio de Janeiro, Rio de Janeiro, Brazil
e-mail: fabio.gatamorta@uerj.br

I. Miskioglu
Michigan Technological University, ME-EM Department, Houghton, MI, USA

E. Bayraktar (✉)
Isae-Supmeca-Paris, School of Mechanical and Manufacturing ENgineering, Paris, France

ISAE-Supmeca, School of Mechanical and Manufacturing Engineering, Paris, France
e-mail: emin.bayraktar@isae-supmeca.fr

F. Gardea et al. (eds.), *Mechanics of Composite, Hybrid and Multifunctional Materials, Volume 5*,
Conference Proceedings of the Society for Experimental Mechanics Series, https://doi.org/10.1007/978-3-031-50478-5_2

(20/1 ball/powder ratio). The chemical composition of the matrix was given as Al-5.5 Zn-2.5 Mg-0.5 Cu; additionally, 4 wt% of zinc stearate was used as a lubricant during the preparation of the composite. After the milling operation, the thermal behaviour of AA431 powder was evaluated using a differential scanning calorimetry thermogravimetric analysis diagram. All of the details for experimental conditions were given in previous papers [10, 12].

This matrix was divided into three parts and reinforced with TiB_2, TiC and B_4C (40 wt%) and as minor reinforcements, molybdenum and copper (Mo 4 wt%, Cu 4 wt%) were used. During the milling, pure nano AA1050 (3–5 wt%) was added to facilitate and homogenize the mixture of two types of recycled aluminium. Sintering was carried out under argon gas in the oven. At the first stage of the sintering, compacted samples were heated at 380 °C for 45 min to eliminate lubricant and other artefacts. At the second stage of sintering, heating was conducted up to the maximum sintering temperature (600 °C). After that, hot forging was carried out to complete the manufacturing processes at a lower temperature than the sintering temperature. Mechanical and physical properties were evaluated through micro-hardness, static-cyclic-time-dependent compression tests and low-velocity impact/drop weight tests were carried out at only one energy level.

Results and Discussion

Figure 1 shows the DSC diagram measuring AA431 and also the simulation of a fraction of solid depending on the temperature calculated using the software "Thermo-Calc" in the matrix to determine the critical transformation points during the heating and cooling stages. Additional information was given for energy-dispersive spectroscopy (EDS) chemical analysis in the next sessions for each composite.

Table 1 presents all three composites formulated for the innovative hybrid composite design. Besides the major reinforcements (TiB_2, TiC, and B_4C) and Mo-Cu, there is a small amount of graphene nanoplatelets (GNPs) that increases, mainly on the grain boundaries. For the fine distribution of the reinforcements and to obtain a fine grain size, Mo, Cu and nanographene were added into the matrix for each composite. Additionally, the presence of Mo GNPs in the structure improved the

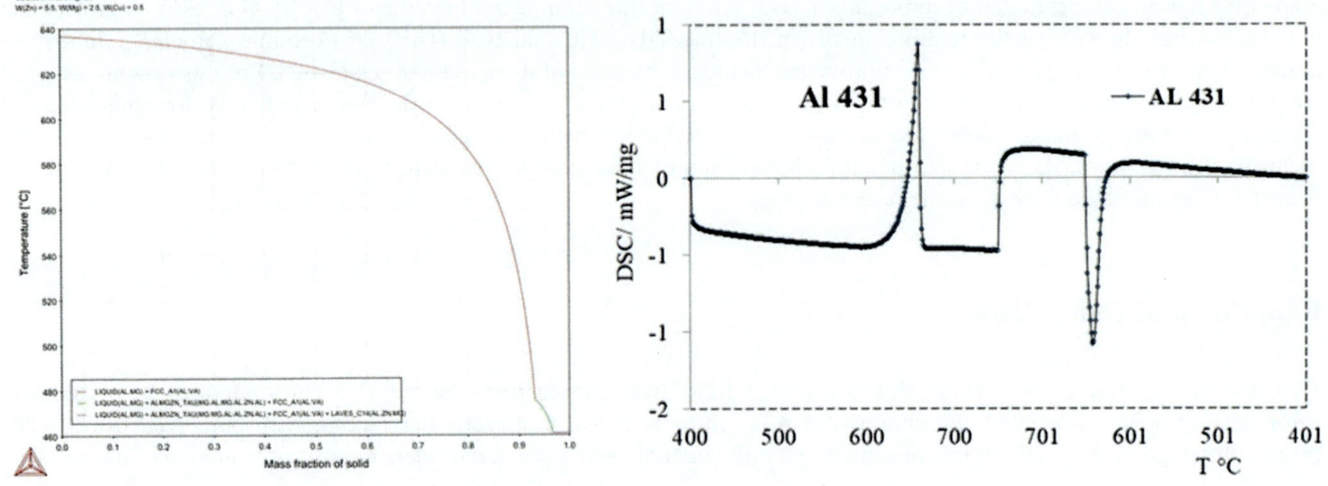

Fig. 1 Differential scanning calorimetry (DSC) diagram measuring AA431 and also simulation of the fraction of solid depending on the temperature [12]

Table 1 Compositions of the three composites (wt%)

Composite name	Al 431 (50%) 1050 (50%) (Matrix)	TiB_2	TiC	B_4C	Mo	Cu	Zn-Stearate	GNPs
E22	B	40			4	4	4	0.25
E23	B		40		4	4	4	0.25
E24	B			40	4	4	4	0.25

Table 2 Electrical and thermal properties measured for three composites with microhardness values

Composites	Electrical conductivity at 20 °C (S/m)	Thermal conductivity (W/mK)	Microhardness ($HV_{0.1}$)
E22	7.05×10^9	6.250	285 ± 15
E23	6.15×10^9	5.356	280 ± 20
E24	5.82×10^9	5130	360 ± 15

mechanical resistance for toughening, mainly because of a strong cohesion at the interface between the matrix and the reinforcements. Cu improves the chemical bonding diffusion in the matrix. These results, which will be presented in the next session, are only indicative and should be improved for our common research project.

Electrical conductivity levels were measured using an "Agilent 4338B Milli-ohm Meter". Four specimens were measured for each composite and then, the mean values were indicated in Table 2.

For the measurements, DC-regulated power supply voltage and current were set as 20 V and 20 A respectively. Data acquisition card "NI9234" was connected in parallel with the output of the power to acquire the voltage data (voltage input accuracy was 24 bits). A high-precision multi-meter "Agilent U1253N" was connected in the series to measure the current intensity (A).

Electrical and thermal conductivity and also microhardness measurements from the samples produced under the same conditions for three composites were shown in Table 2.

All of the data for the electrical and thermal measurements were revealed using the LabVIEW program. These results should obviously be assumed to be indicative data.

Microstructural evolution has been perfectly carried out on the SEM with EDS chemical analyses. A strong cohesion between the matrix and reinforcements was observed on the specimens for three composites.

Figure 2 gives certain results from general microstructure and mapping; detection of the TiB_2 in the composite containing 40wt% TiB_2 respectively. The fine particles of TiB_2 can easily give a very sound cohesion with matrix and very often chemical bonding diffusion accelerated with ultra-fine copper particles. However, some floccule agglomeration in certain specimens was found under laboratory conditions. Evidently, a long ball milling is needed for a homogenous mixture.

Figure 3 gives certain results from general microstructure and mapping; detection of the TiC in the composite containing 40wt% TiC respectively. The fine particles of TiC can be easily found with much more homogenous distribution in the matrix giving a very sound cohesion and chemical bonding diffusion was also very easily observed, again accelerated with ultra-fine copper particles.

Figure 4 also gives the results from general microstructure and mapping and detection of the B_4C in the composite containing 40wt% B_4C respectively. The fine particles of B_4C have shown a homogenous distribution in the matrix, giving a very sound cohesion. Much more homogenous chemical bonding diffusion was also observed, again accelerated with ultra-fine copper particles.

Figure 5 gives a general presentation of the manufacturing process on the hydraulic press carried out at 600 °C followed by hot forging of the specimens. More details of the compression test device on the Zwick test machine (ISAE-SUPMECA/Paris) and also damaged specimens with 45°cracks (max shear stress) after the compression test were also presented in this figure. Machining chips for three composites were also given here to show easy machining of the composites. High surface qualities of the machined specimens were obtained. Even if there were high ceramic-based reinforcements in the aluminium matrix, any brittleness was observed during the machining process.

Static and incremental cyclic compression test results were performed to evaluate the mechanical behaviours of these innovative hybrid composites, with a deformation rate of 1.5 mm/min. A Zwick test machine (ISAE-SUPMECA/Paris) with a load cell capacity of 100 kN was used for the static and incremental compression tests.

These graphs have been presented for three composites.

Low cycle fatigue (LCF-Oligocycle) tests have been performed closed loop test device with a frequency of 0.5HZ. LCF tests were performed with a computer adapted with 100-kN load cell (ASTM E606). These tests were conducted under fully reversed compression–compression loading. LCF tests were limited as 10^4 cycles and the maximum strength was chosen to be at around the yield strength, between 70% and 80% of the maximum strength.

As indicated in the former section, the cyclic deformation behaviour of metal matrix composites is fully related to the homogeneity of the composite microstructure, mainly the distribution and strong cohesion of the reinforcements with the matrix interface. Here, a mixture of the two recycled forms of aluminium (Alumix A431 and pure A1050) is used as the matrix material to compare the effects of fine reinforcements on the LCF behaviour of these three composites.

Figures 6, 7 and 8 show the total results of the mechanical behaviour of the three composites under the static incremental cycle and time depending on LCF conditions respectively.

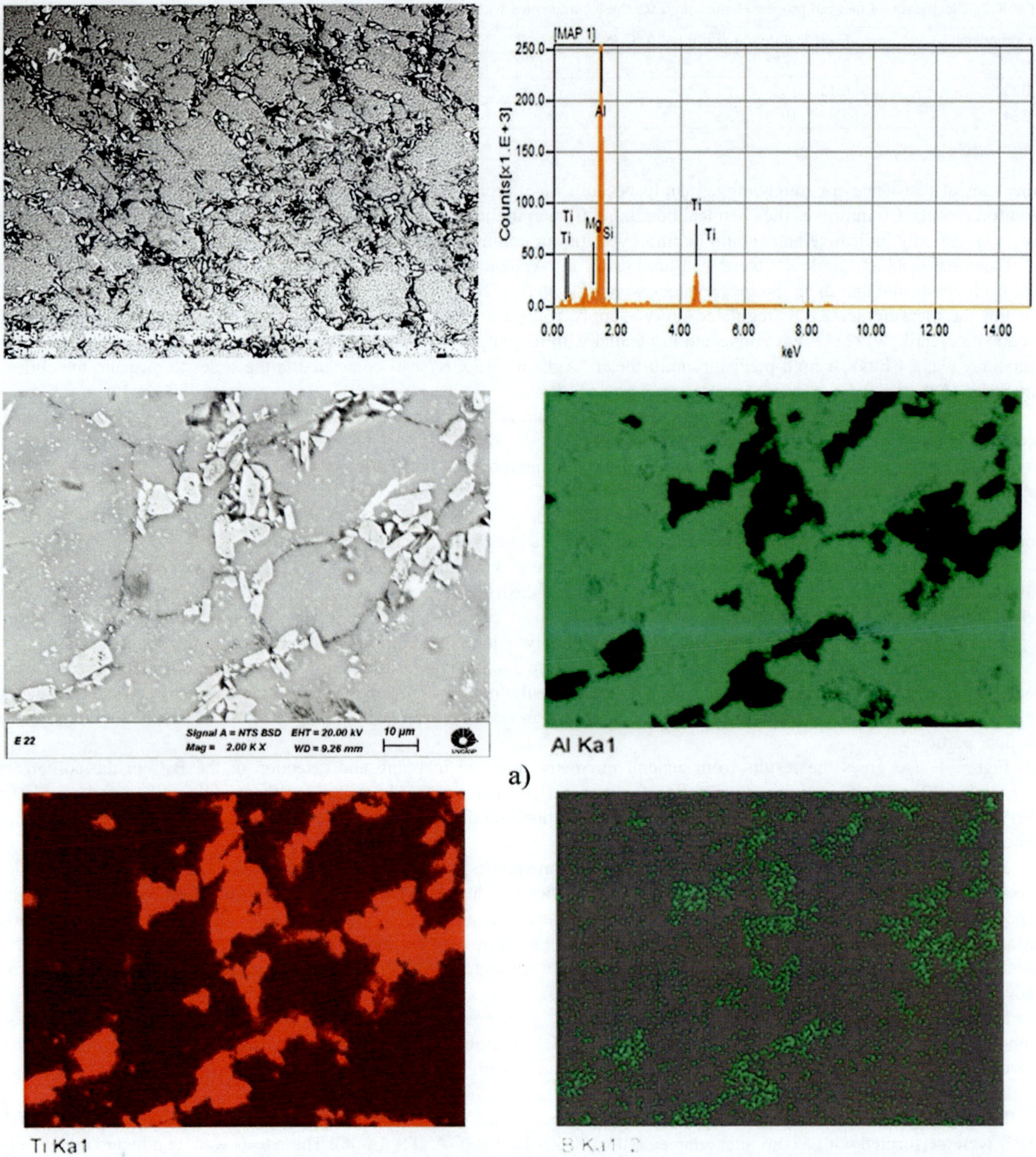

Fig. 2 (**a**) Microstructure of the composite (E22) containing 40wt% TiB2 and (**b**) mapping; detection of the reinforcements at the grain boundary around TiB_2

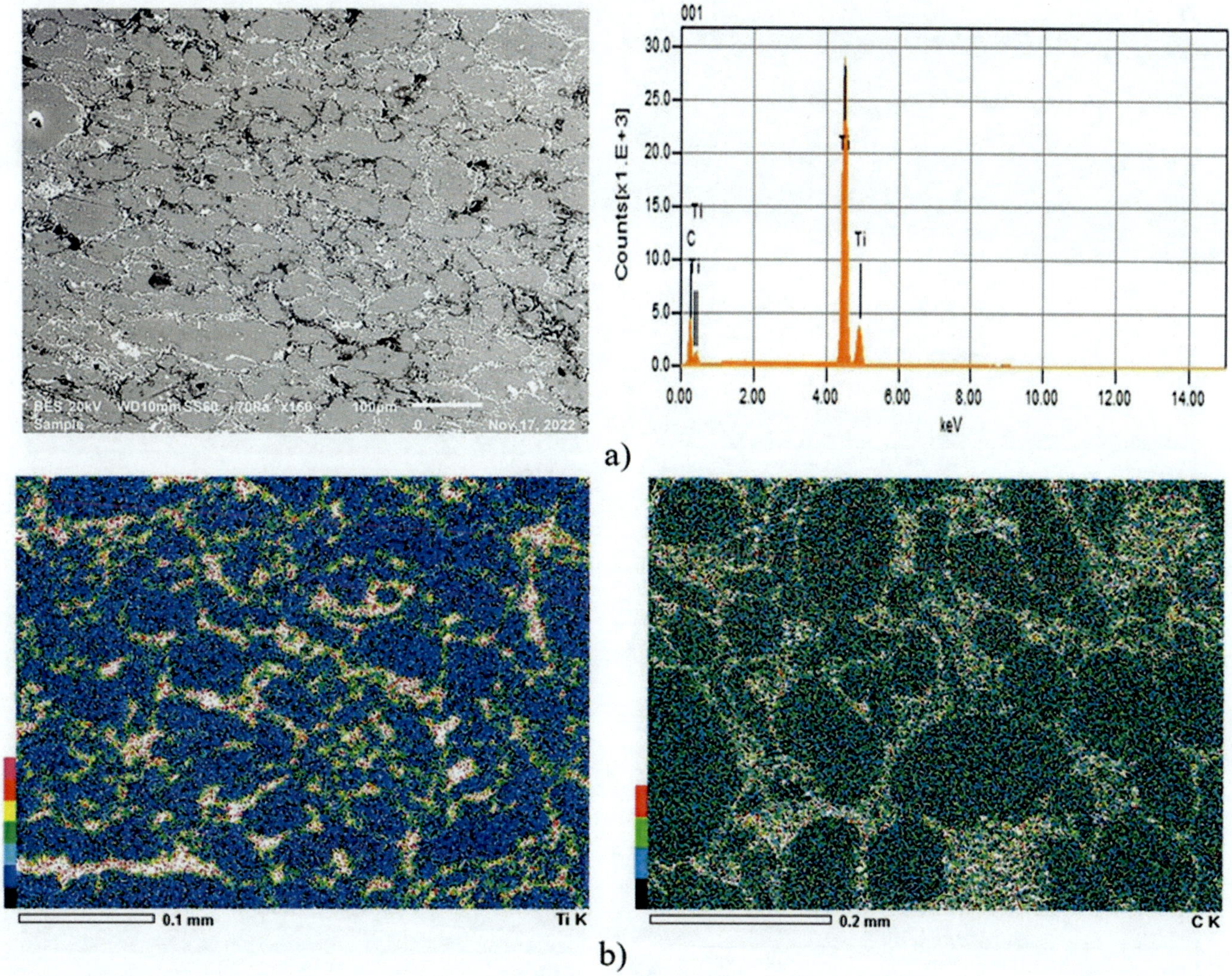

Fig. 3 (a) Microstructure of the composite (E23) containing 40wt% TiC and (b) mapping; detection of the TiC reinforcements in the matrix with a homogenous distribution

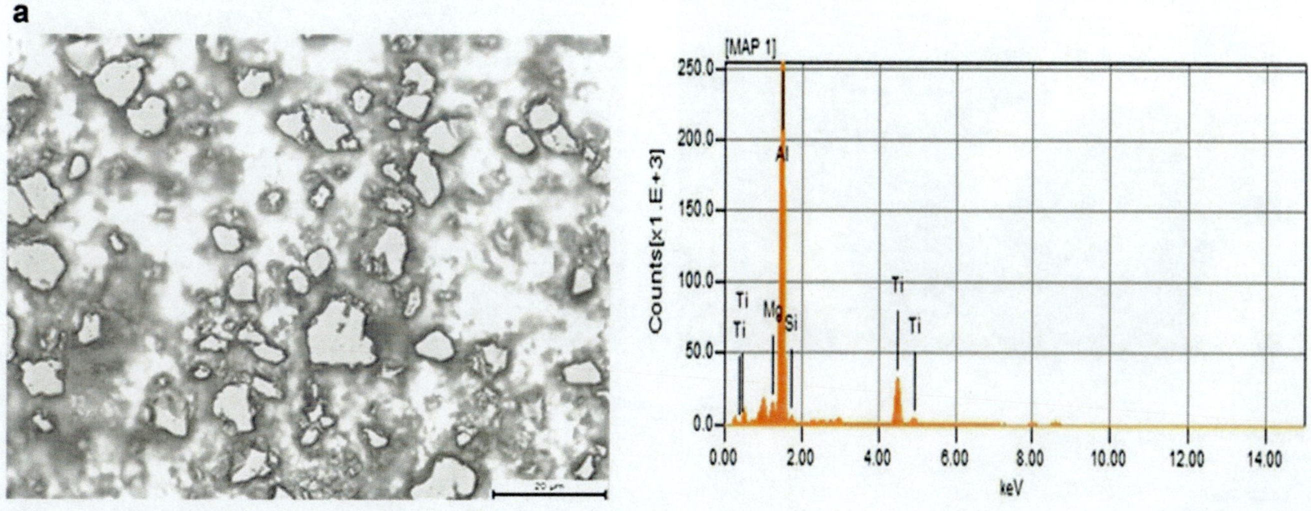

Fig. 4 (**a**) Microstructure of the composite containing 40 wt% B_4C and EDS chemical analyses. (**b**) Mapping of the microstructure of the composite (E24) containing 40 wt% B_4C and detection of the B_4C reinforcements in the matrix with a homogenous distribution

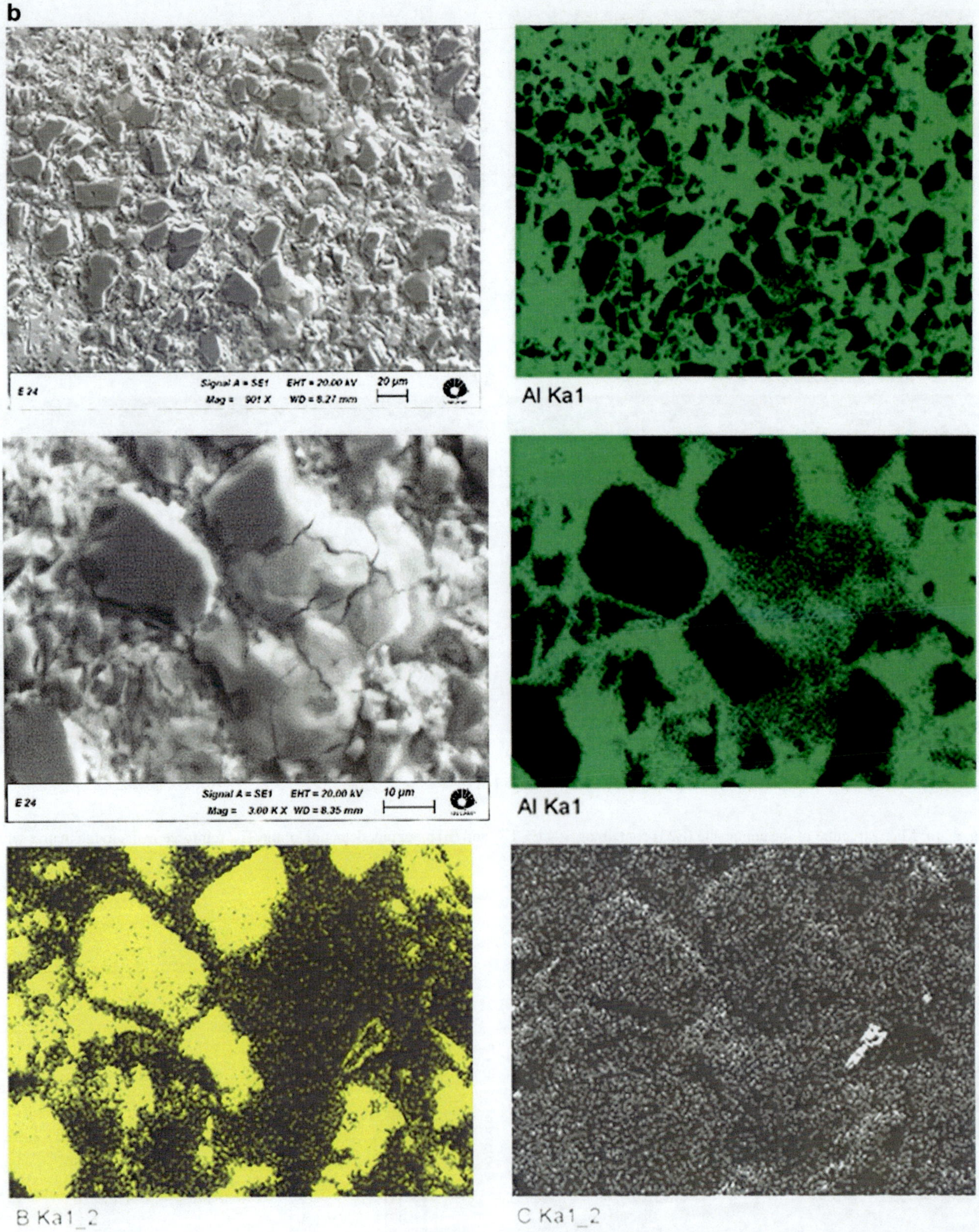

Fig. 4 (continued)

Fig. 5 Sintering at 600 °C followed by hot forging of the specimens (left), a compression test device on the Zwick test machine (ISAE-SUPMECA/Paris) and also damaged specimens with 45°cracks (maximum shear stress) after the compression test. Machining chips for three composites after the manufacturing of the specimens

A special analysis was carried out, only for detection of the fatigue crack propagation in the composite E22 containing 40wt% TiB2 (Fig. 9). These analyses were also made on the fatigue damage specimens for two other composites in order to observe the fatigue damage for each composite.

All of the fatigue specimens produced from three composite have shown transgranular fatigue damage behaviour, most of them in the matrix that has relatively large grains. The cohesion of the reinforcements with the matrix are very strong and sound chemical bonding diffusion was observed each time. Figure 9 was given here only as an example to show this type of behaviour. Additionally, some of the research results published in the literature said that the LCF life in metal matrix composites is generally shorter than that of the unreinforced matrix [4, 6–9, 11, 13]. They have also indicated that the integration of ceramic particle reinforcements into the metal matrix can decrease the ductility and toughness of the metal matrix composites extensively.

In the present work, all of the specimens tested for three composites exhibit ductile fracture morphology. At the fracture surfaces, the reinforcements keep their position with a sound cohesion with the matrix. The first composite containing 40 wt% TiB_2 composite displayed quite a stable cyclic reaction at the same strain levels, which are around 0.7–0.8%. These values are higher than those found in the literature [12, 15]. The second composite containing 40 wt% TiC and the third composite containing 40 wt% B_4C have shown a very slight cyclic hardening followed by softening at the same strain levels of 0.7–0.8%. These results have been carried out under laboratory conditions; they should be improved by another repeated test to obtain safe aircraft engineering applications.

As for the mean values for the mechanical test results, Table 3 show the results of static compression tests and Table 4 indicates the fatigue test results. All of the results were indicated as the mean values.

It should be concluded that the three composites show more ductile behaviour, which is related to the tough and sound microstructure with a strong cohesion of the reinforcements with the matrix. There is a strong trend on the influence of the deformation rates; lower deformation rates give higher resistance, probably related to the recovery of the microstructure at the interface of the reinforcement/matrix.

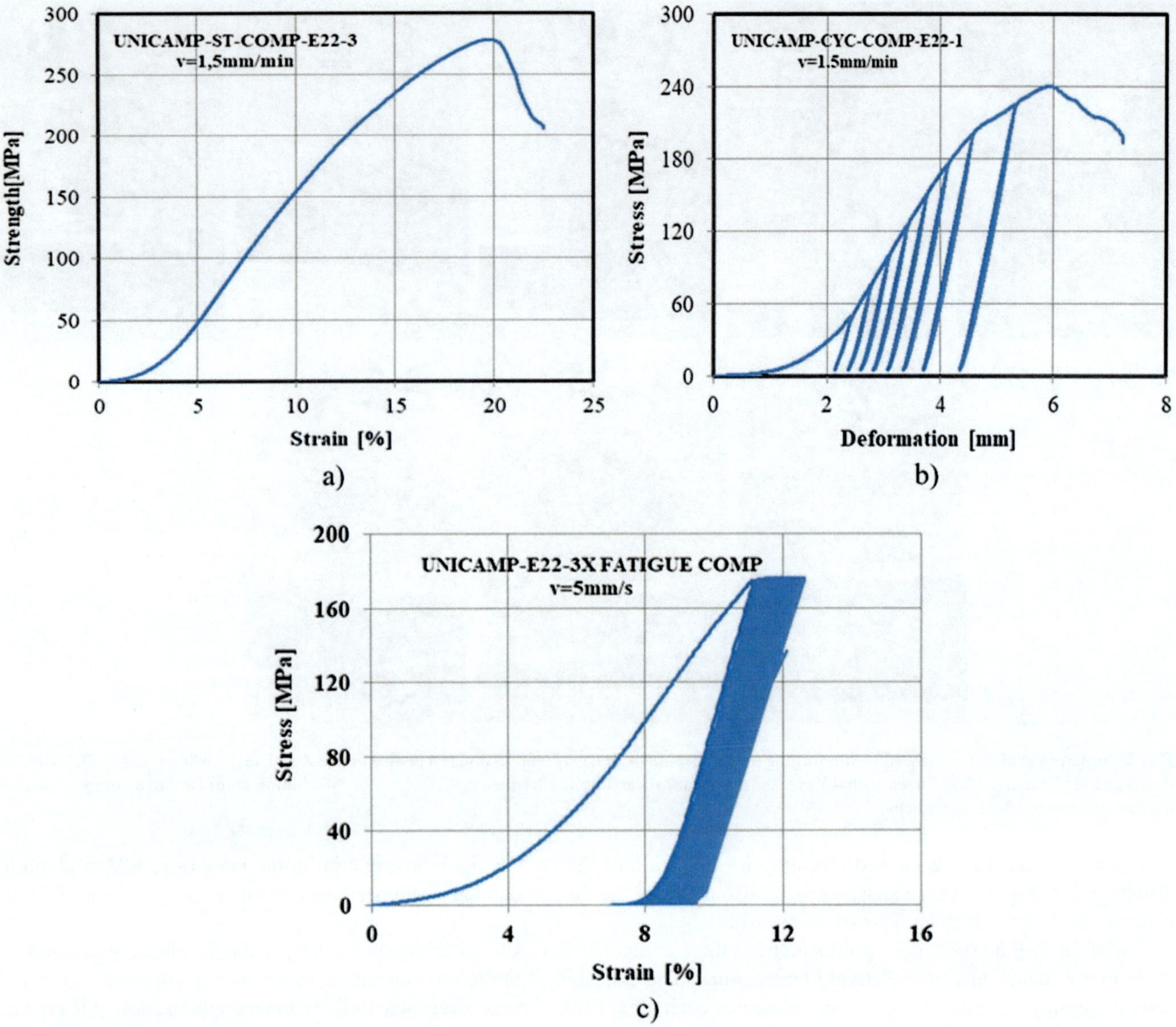

Fig. 6 Mechanical behaviours of the composite (E2) containing 40 wt% TiB2 in the case of compression tests: static (**a**), incremental cyclic (**b**) and fatigue behaviours (**c**). Only for sintered specimens

Conclusions

Innovative aluminium (Al 431 + A1050) matrix composites were designed from the recycled chips supplied by a Brazilian aeronautic company, using the combined method of powder metallurgy followed by "sintering + forging". In the present work, three different composites were designed. As the primary reinforcements, fine ceramic powders defined as 40 wt% TiB_2, TiC and B_4C were used and as the secondary reinforcements based on Cu/Mo/GNPs/γ-Al_2O_3 fibre were used as a low-cost manufacturing process for use in aeronautical engineering applications. In a special case for the present work, some of the applications were given for certain pieces of thermocompressors.

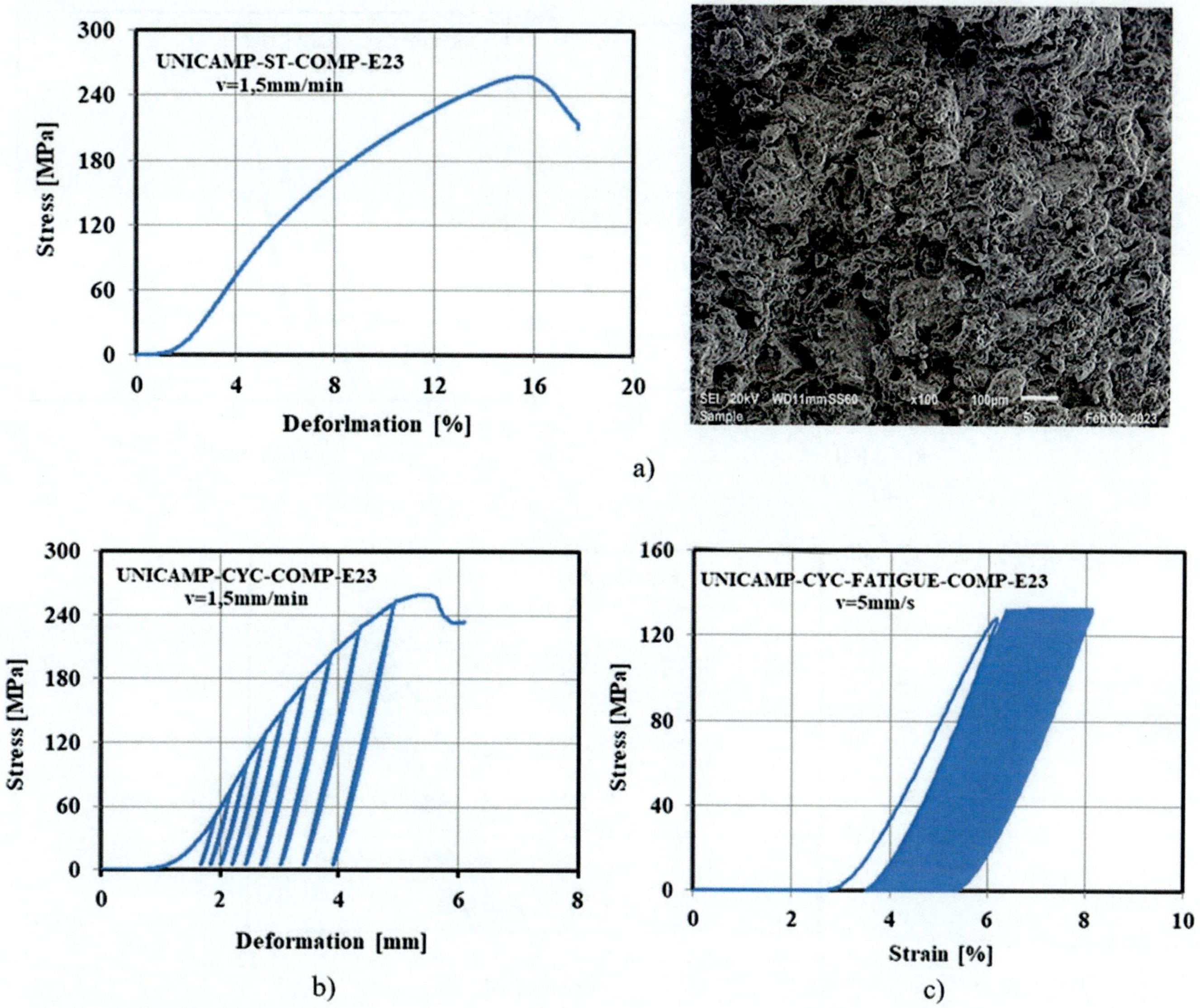

Fig. 7 Mechanical behaviours of the composite (E23) containing 40 wt% TiC in the case of compression tests: static test and fracture surface (**a**), incremental cyclic (**b**) and fatigue behaviours (**c**). Only for sintered specimens

Generally, the static and incremental cyclic compression tests and also the LCF tests give promising results. For the next stage of this common research project, operational parameters for the static and cyclic tests should be taken into account to optimize the composition with the reinforcement levels.

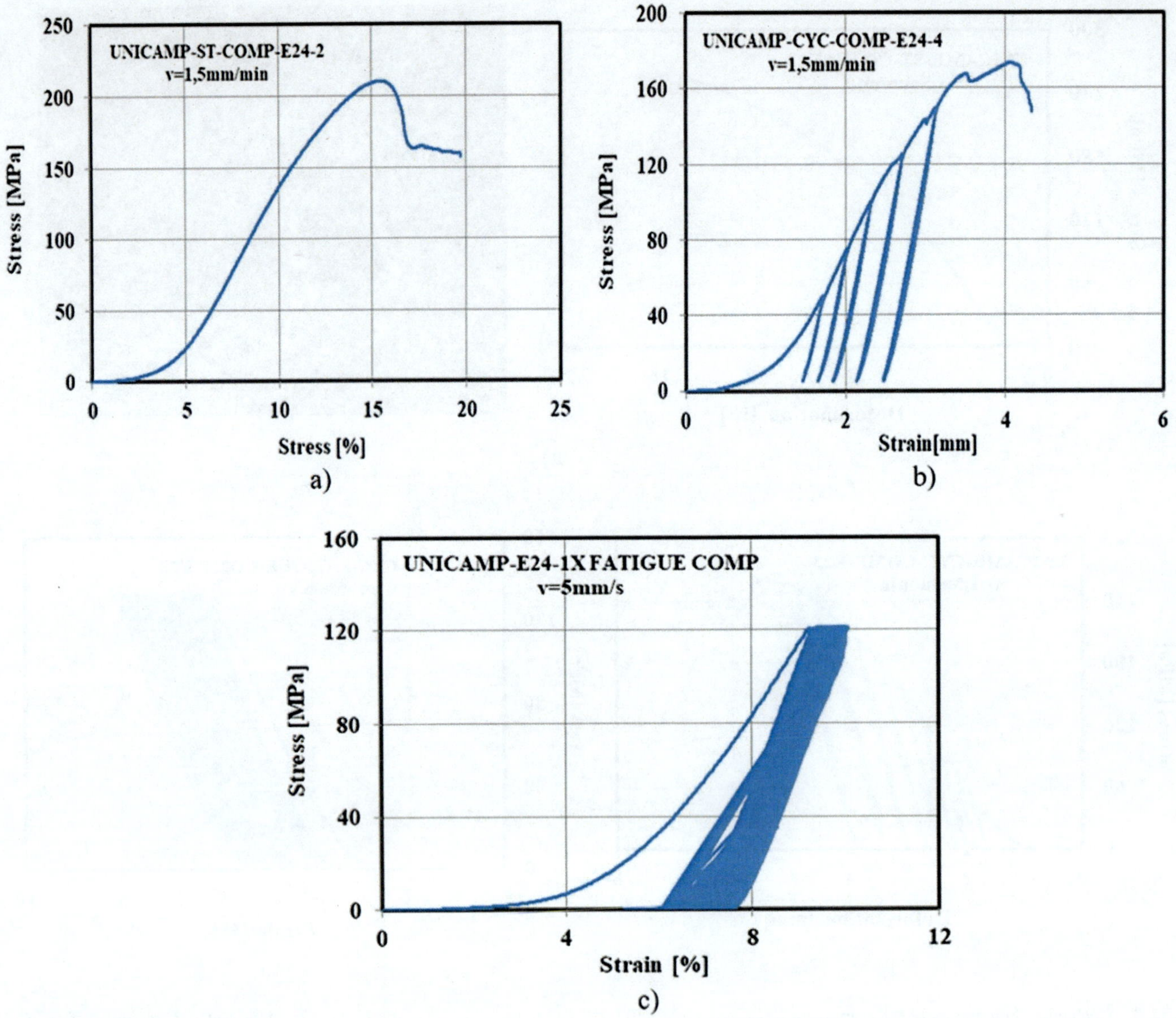

Fig. 8 Mechanical behaviours of the composite (E24) containing 40wt% B4C in the case of compression tests: static (**a**), incremental cyclic (**b**) and fatigue tests (**c**). Only for sintered specimens

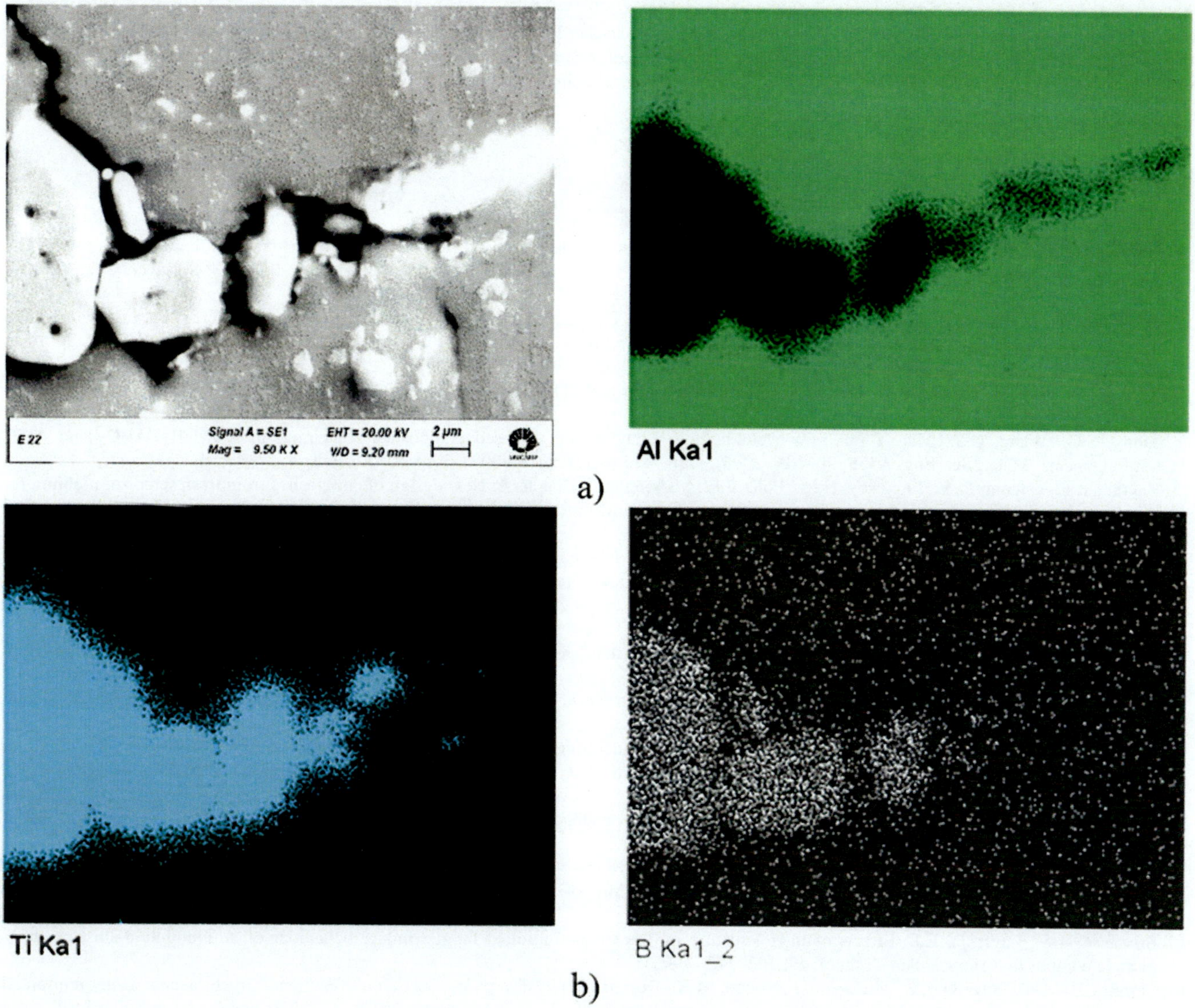

Fig. 9 Mapping in the composite (E22) containing 40wt% TiB_2 (**a**) and detection of the TiB_2 around the intergranular fatigue crack propagation in the matrix (**b**)

Table 3 Static compression test results for the three composites given for the sintered and sintered+ forged specimens

Composite name	UTS (MPa) Sintering (mean values)	UTS (MPa) Sinter + forging (not broken)
E22 – 40 wt% TiB_2	280	560
E23 – 40 wt% TiC	275	535
E24 – 40 wt% B_4C	255	570

Table 4 Fatigue compression test results of the three composites given for the sintered and sintered + forged specimens

Maximum fatigue stress (mean values) without failure at 10,000 cycles	E22 40 wt% TiB_2 σ_{max} (MPa)	E23 40 wt% TiC σ_{max} (MPa)	E24 40 wt% B_4C σ_{max} (MPa)
Sintering	220	210	215
Sinter + forging	430	415	450

Acknowledgements Authors acknowledge financial support from CNPq – National Council for Scientific and Technological Development, (Brazil) and Program French Cathedra UNICAMP/French Consulate in São Paulo, Brazil. The third author, E. Bayraktar, personally thanks the French Consulate in Sao Paulo for partial financial support in this research project. We acknowledge Professor Dr R. Caram, head of the physical metallurgy laboratory, UNICAMP/FEM – Campinas, Brazil for technical help and valuable discussion of the results.

References

1. Ferreira, L.F.P., Bayraktar, E., Miskioglu, I., Katundi, D.: Chapter 15: Design of hybrid composites from scrap aluminium bronze chips. In: Mechanics of Composite and Multi-functional Materials Society for Experimental Mechanics Series, pp. 131–138. Springer, Cham (2016). ISBN 978-3-319-41766-0
2. Gatamorta, F., Larbi, A., Bayraktar, E., Miskioglu, I.: Chapter 7: Design of copper and silicon carbide (SiC) reinforced recycled aluminium matrix composites through sintering + forging. In: Mechanics of Composite, Hybrid and Multifunctional Materials, vol. 6, pp. 45–52. Springer, Cham (2020). ISBN 978-3-030-59867-9, https://doi.org/10.1007/978-3-030-59868-6-7
3. Rodrigo, P., Poza, P., Utrilla, M.V., Ureña, A.: Identification of σ and Ω phases in AA2009/SiC composites. J Alloys Compd. **482**, 187–195 (2009)
4. Tjong, S.C., Wang, G.S., Mai, Y.W.: Low-cycle fatigue behaviour of Al-based composites containing in situ TiB_2, Al_2O_3 and Al_3Ti reinforcements. Mater. Sci. Eng. **A358**, 99–106 (2003). https://doi.org/10.1016/S0921-5093(03)00266-1
5. Bayraktar, E., Gatamorta, F., Enginsoy, H.M., Polis, J.E., Miskioglu, I.: Chapter 6: New design of composites from fresh scraps of niobium for tribological applications. In: Mechanics of Composite, Hybrid and Multifunctional Materials, vol. 6, pp. 35–43. Springer, Cham (2020). ISBN 978-3-030-59867-9, https://doi.org/10.1007/978-3-030-59868-6-6
6. Popov, V.A., Shelekhov, E.V., Prosviryakov, A.S., Kotov, A.D., Khomutov, M.G.: Particulate metal matrix composites development on the basis of in situ synthesis of TiC reinforcing nanoparticles during mechanical alloying. J Alloys Compd. **707**, 365–370 (2017)
7. Luk, M.J., Mirza, F.A., Chen, D.L., Ni, D.R., Xiao, B.L., Ma, Z.Y.: Low cycle fatigue of SiCp reinforced AA2009 composites. Mater. Des. **66**, 274–283 (2015)
8. Hadianfard, M.J., Mai, Y.W.: Low cycle fatigue behaviour of particulate reinforced metal matrix composites. J. Mater. Sci. **35**, 1715–1723 (2000)
9. Srivatsan, T.S., Annigeri, R.: The quasi-static and cyclic fatigue fracture behavior of 2014 aluminum alloy metal-matrix composites. Metall. Mater. Trans. A. **31**, 959–974 (2000)
10. Ferreira, L.F.P., Gatamorta, F., Bayraktar, E., Robert, M.H.: Manufacturing of low cost composites with porous structures from scrap aluminium (AA2014) chips. In: Mechanics of Composite and Multi-functional Materials, vol. 7, pp. 233–240. Springer, Cham (2017). https://doi.org/10.1007/978-3-319-41766-0_28
11. Zhang, Q., Chen, D.L.: A model for predicting the particle size dependence of the low cycle fatigue life in discontinuously reinforced MMCs. Scr. Mater. **51**(9), 863–867 (2004)
12. Gatamorta, F., Miskioglu, I., Bayraktar, E., Melo, M.L.N.M.: Recycling of aluminium-431 by high energy milling reinforced with TiC-Mo-Cu for new composites in connection applications. In: Mechanics of Composite and Multi-functional Materials, vol. 5, pp. 41–46. Springer, Cham (2019). https://doi.org/10.1007/978-3-030-30028-9_6
13. Srivatsan, T.S., Lavernia, E.J.: Effects of microstructure on the strain-controlled fatigue failure behaviour of an aluminium-alloy/ceramic-particle composite. Compos. Sci. Technol. **49**, 303–313 (1993)
14. Ferreira, L.-M.-P., Bayraktar, E., Miskioglu, I., Robert, M.H.: Design and physical properties of multifunctional structural composites reinforced with nanoparticles for aeronautical applications. Adv. Mater. Process. Technol. **3**(1), 33–44 (2017). Taylor & Francis-USA
15. Gatamorta, F., Miskioglu, I., Bayraktar, E., Melo, M.L.N.M.: Recycling of aluminium-431 by high energy milling reinforced with TiC-Mo-Cu for new composites in connection applications. In: Mechanics of Composite and Multi-functional Materials, vol. 5, pp. 41–46. Springer (2019). https://doi.org/10.1007/978-3-030-30028-9_6

Static and Fatigue Behaviour of Recycled Thin Sheet "Ti-Al-Nb" Based Composites Produced by Hot Forging Diffusion Process

Georges Zambelis, Fabio Gatamorta, Özgür Aslan, Ibrahim Miskioglu, and Emin Bayraktar

Abstract Within the framework of the common research project, the mechanical properties and fatigue behaviour of recycled thin sheet Ti-Al-based composites reinforced with atomized scrap aluminium (AA7075) and Nb elements have been evaluated. All the thin sheet sandwich structures were produced by the hot forging process, which is a semi-solid-forming process similar to partial melting hot forging. The effect of the chemical bonds during the production of these multifunctional sandwich composite structures was analysed using 3-point bending tests under static and dynamic (fatigue) loading conditions. Additional tensile tests have been carried out to evaluate the mating effect. Interface and microstructure of these composites have also been evaluated using scanning electron microscopy.

Keywords Recycled Ti-Al · SEM · 3PB · Static · Tensile test

Introduction

The development of TiAl intermetallics is a practical solution for the aeronautical and/or aerospace engineering owing to their exceptional properties such as low density, high stiffness, high strength, high resistance to corrosion/oxidation etc. In general, TiAl, Ni-Al Nb_2Al intermetallics are structural materials that are being considered as ideal new high-temperature structural materials for civil and military applications [1, 2]. To alleviate safety concerns in the aerospace industry, the application of TiAl intermetallics needs a reliable manufacturing process such as diffusion bonding, with different materials to construct a new composite family [3–10].

Owing to the excellent mechanical properties of TiAl intermetallics-based composites with a healthy and sound microstructure, the new design of these composites was developed within the framework of a joint research project with the French Aeronautical Society. A novel hot forging process/diffusion bonding followed by a second heat treatment to reduce the residual stresses and attain a relatively soft and ductile structure. A chemical diffusion bonding process was carried out at the interface between the two thin sheets of the composite. The present work shows a low-cost reliable manufacturing process for easy chemical diffusion bonding for the new intermetallics-based composites. Detailed experimental tests were carried out to evaluate the static and cyclic, time-dependent properties of these composites. Microstructural analyses were carried out using scanning electron microscopy (SEM).

G. Zambelis
SAFRAN GROUP – Innovation Composites Research Centre, Paris, France

F. Gatamorta
State University of Rio de Janeiro, Rio de Janeiro, Brazil

Ö. Aslan
Atilim University, Department of Computational Mechanics, Ankara, Turkey

I. Miskioglu (✉)
Michigan Technological University, ME-EM Department, Houghton, MI, USA
e-mail: imiski@mtu.edu

E. Bayraktar
UNICAMP, University of Campinas, FEM, Campinas, SP, Brazil

ISAE-Supmeca, School of Mechanical and Manufacturing Engineering, Paris, France

F. Gardea et al. (eds.), *Mechanics of Composite, Hybrid and Multifunctional Materials, Volume 5*,
Conference Proceedings of the Society for Experimental Mechanics Series, https://doi.org/10.1007/978-3-031-50478-5_3

Experimental Conditions

As a practical manufacturing process, three recycled Ti-AA7075 titanium aluminium intermetallic sheets containing 10 wt% fine niobium powder were prepared as a sandwich structure. The surface roughness of the thin sheets was increased by polishing them with a hard wire steel brush and a steel needle. After the thin sheets had been cleaned, they were prepared as a sandwich structure using a hot forging bonding process at 650 °C for around 1 h under a pressure of 200 MPa. The thermal behaviour of the AA7075 aluminium alloy was evaluated with differential scanning calorimetry thermogravimetric analysis (DSC-TGA) and X-ray diffraction (XRD) diagrams for the composite. More details analyses of the experimental conditions were given in the previously published papers [1, 3, 6, 7, 9].

At least ten sandwich structures were handled using the hot forging bonding process at the laboratory scale for the optimization of the key parameters to give a realistic recipe for the industrial scale manufacturing process. Mechanical and physical properties were evaluated through micro-hardness, static-cyclic compression tests, three-point bending (3PB) test as static and cyclic fatigue conditions to evaluate stiffness for the time-dependent behaviour of these novel processes. SEM was used for mapping analyses at the interface of the sandwich structure before and after static 3PB tests.

Results and Discussion

Figure 1a shows the XRD diagram for Ti-AA7075 intermetallics-based composites indicating the phases and additional information was given by energy-dispersive spectroscopy (EDS) chemical analysis for the composite used here. Figure 1b presents a DSC diagram for the AA7075 alloy and simulation of the fraction of solid depending on the temperature calculated using the software "Thermo-Calc" in the matrix to determine the critical transformation points during the heating and cooling stages.

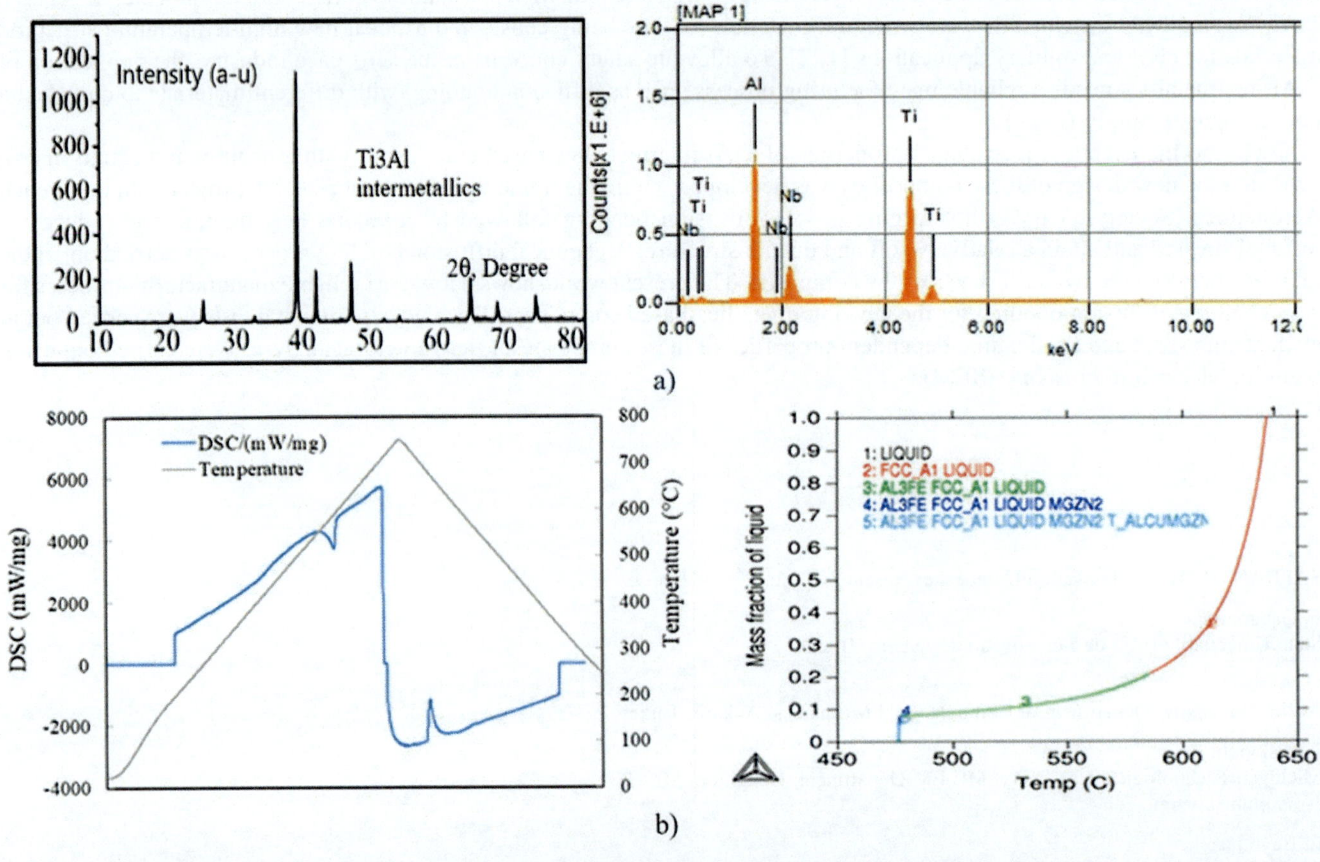

Fig. 1 (**a**) XRD diagram for Ti-AA7075 intermetallics-based composites indicating the phases and additional information was given by EDS chemical analysis for the composite used here. (**b**) DSC diagram measured for the A7075 alloy with a heating rate of 5 °C/min and simulation of the fraction of solid depending on the temperature for A7075 [3, 6, 9]

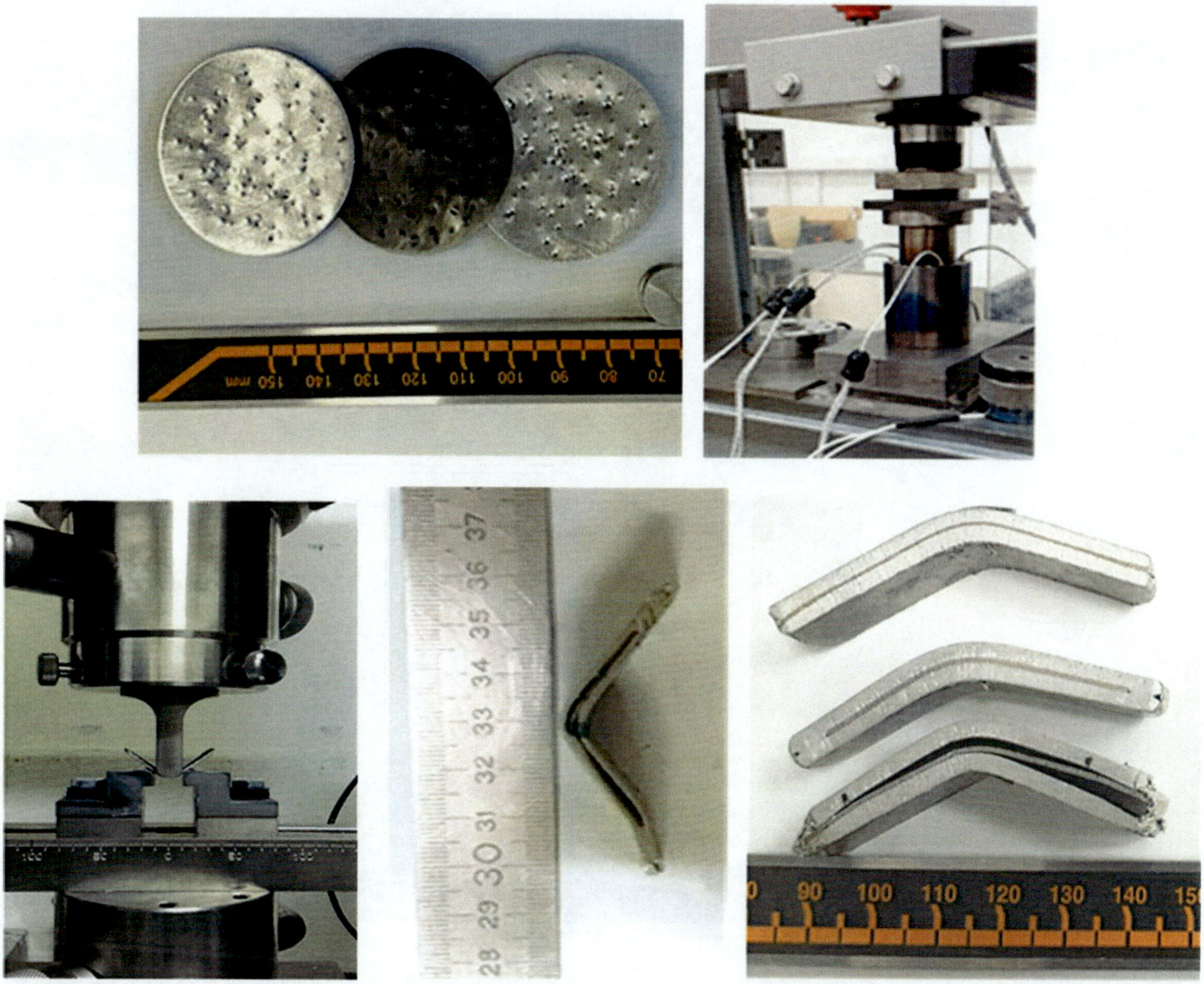

Fig. 2 Hot forging diffusion bonding process carried out at 650 °C adapted on the Zwick test machine and specimens tested under static 3PB devices together on the test machine (ISAE-SUPMECA/Paris)

It seems that XRD patterns of the surface of the composite mainly justifies an intermetallic Ti_3Al phase, which is supported by the EDS analyses of the sandwich sheet structure.

Figure 2 summarizes the manufacturing process of the sandwich structure; thin sheets were subjected to the hot forging bonding diffusion process adapted on the test machine and static 3PB system adapted on the Zwick test device carried out at the composite laboratory. This novel process replaces conventional manufacturing processes and is a simple and efficient way of joining different thin sheets at high temperature under pressure to bring about a diffusion bonding phenomenon between the sheets (Fig. 3). Many sandwich structures obtained this way have shown a healthy microstructure and a high stress level. As will be seen in the next section, these sandwich structures were also tested under tensile test conditions to evaluate the mating effect on the mechanical behaviour of the sandwich structure.

Very strong cohesion can be observed between the thin sheets before and after the static 3PB test (Fig. 3). These mappings were taken on SEM.

To evaluate the mechanical behaviour of the sandwich structures of the Ti-A7075-Nb intermetallics-based composites, three different mechanical test series were conducted on the specimens prepared for each test type. Static 3PB tests, time-dependent cyclic (fatigue) 3PB tests and finally tensile tests were carried out. Five samples were tested to obtain an optimal mean value. Figure 4 shows a typical static 3PB test graph obtained on these specimens. Finally, the maximum stress and yield stress were obtained according to the equations defined by the American Society for Testing and Materials standard.

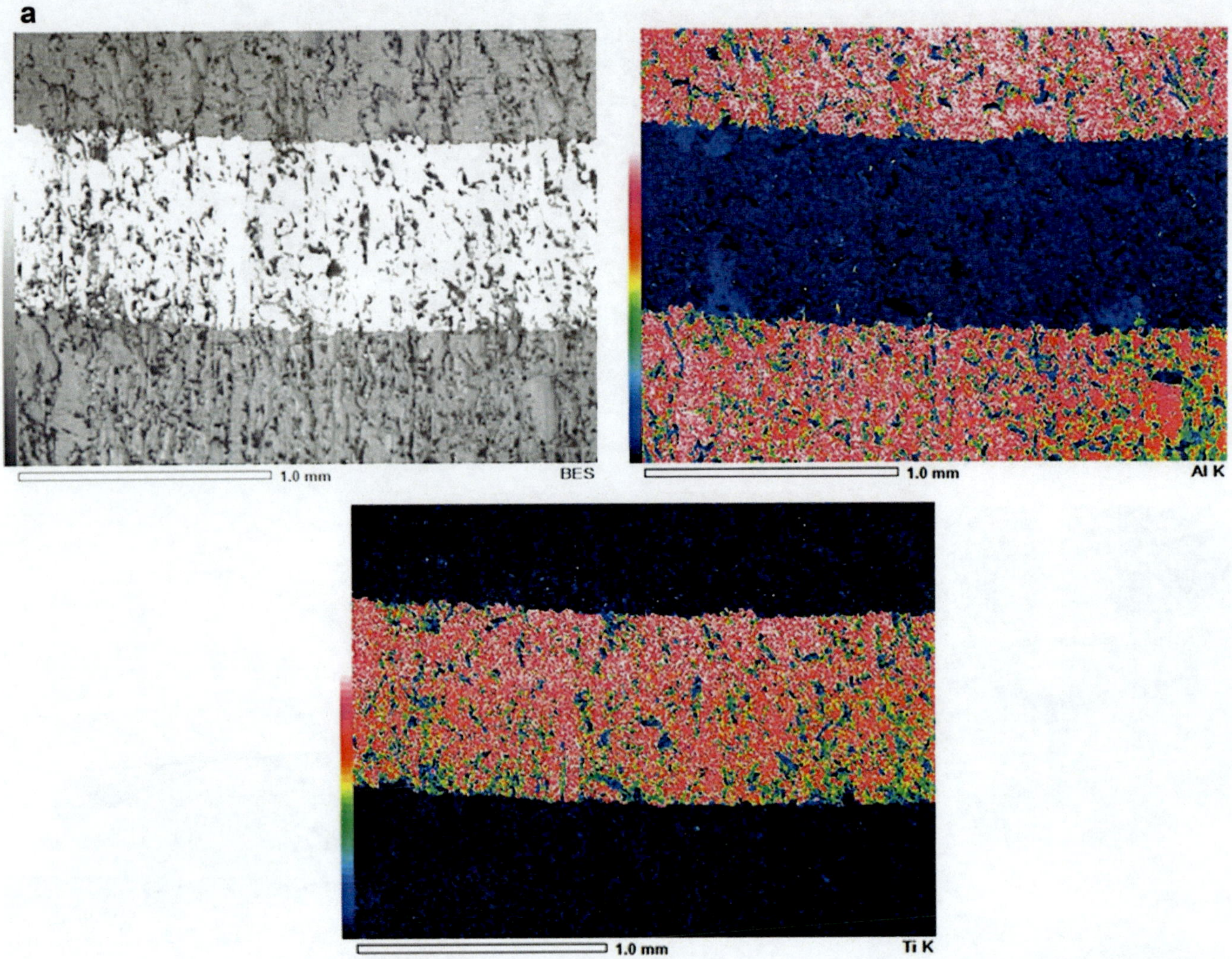

Fig. 3 (**a**) Comparative study for the microstructure of the sandwich structure of the composite Ti-AA7075 mapping analyses before the static 3PB test obtained on SEM. (**b**) Comparative study for the microstructure of the sandwich structure of the composite Ti-AA7075 mapping analyses after the static 3PB test obtained on SEM

As we have indicated before, maximum stress and maximum yield during the deformation were calculated using the following equations:

$$\sigma = \frac{3 \times F \times L}{b \times t^2}$$

and

$$y = \frac{F \times L^3}{4 \times E \times b \times t^3},$$

where F is the applied force, L is the distance between two support points, b is the length of the specimen, t is the thickness of the specimen and E is Young's modulus.

The mean values obtained from these tests are variable depending on the specimen's size from 200 MPa up to 500 MPa; certain specimens have shown local decohesion and/or delamination.

A numerical correlation study has been conducted for the 3PB test. Comparison of the numerical correlation and experiments is shown in Fig. 5 for only one sandwich structure series (AA7075-Ti-Nb) with the specimen no. 1. Figure 5a shows the maximum stress vs strain and Fig. 5b shows numerical correlation of the same specimen with force displacement graph.

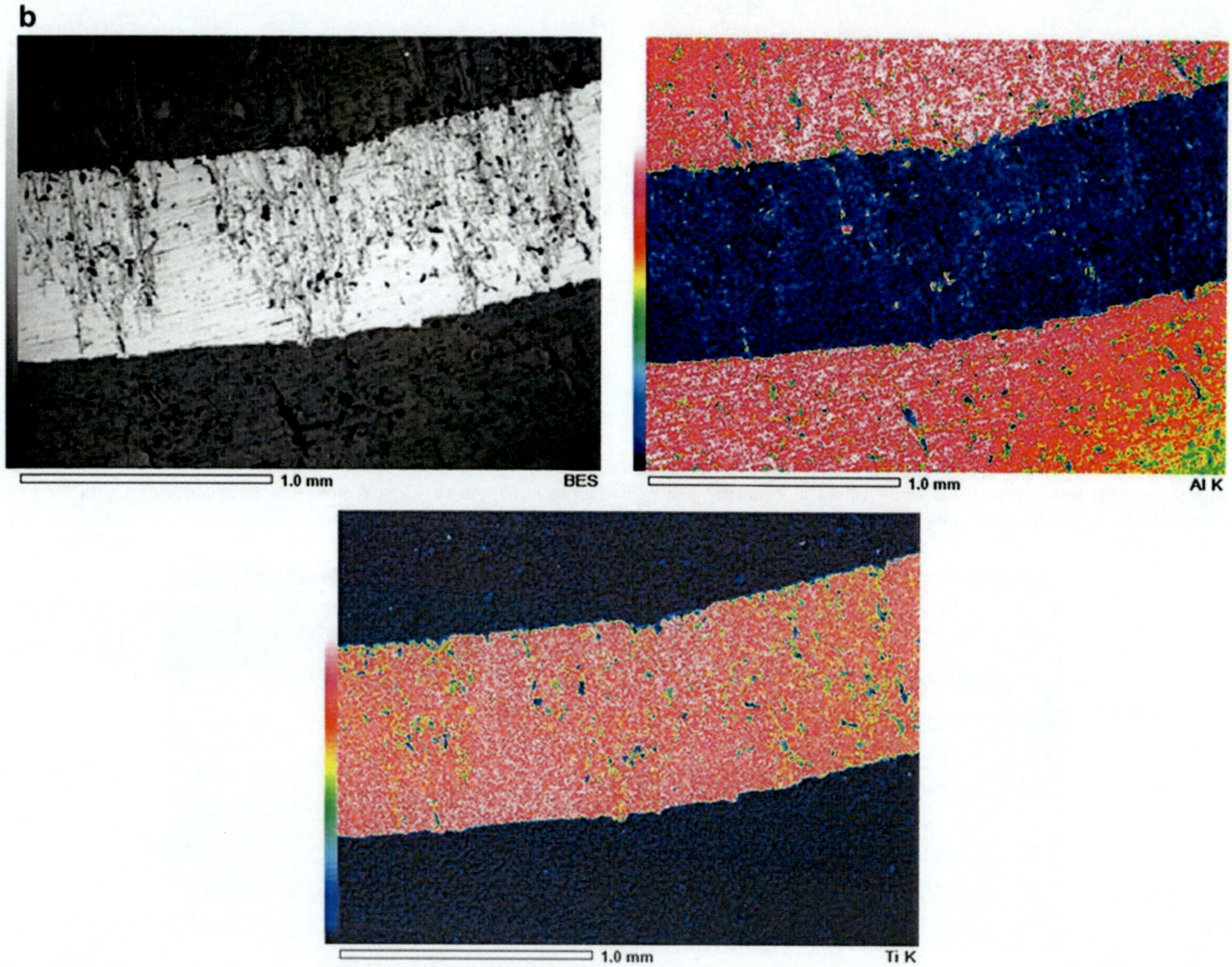

Fig. 3 (continued)

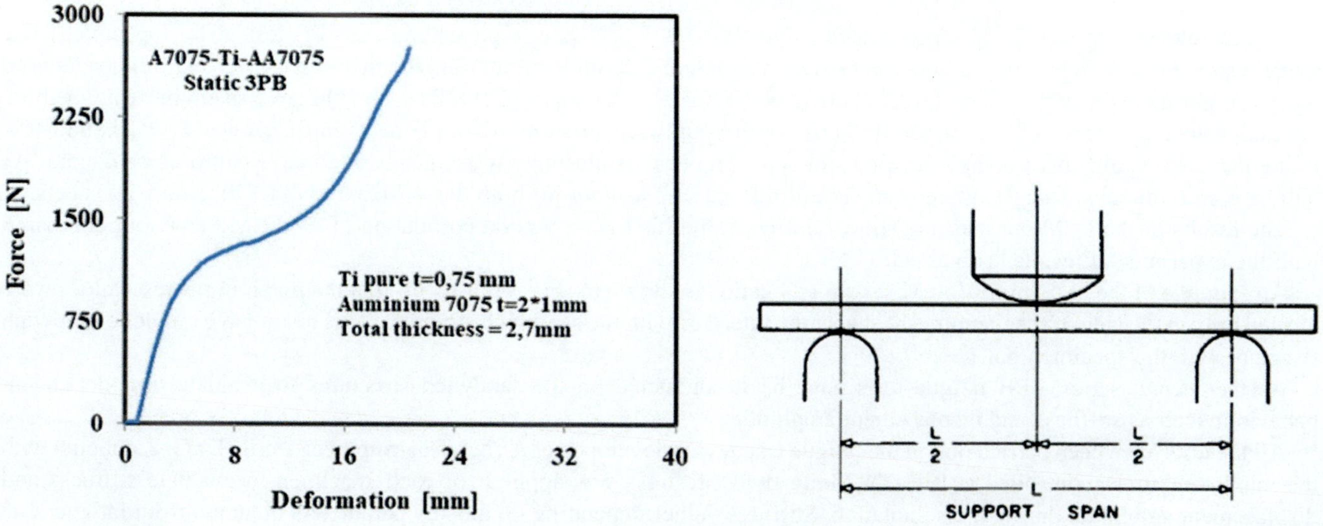

Fig. 4 Static three-point bending test carried out on the sandwich structure (Ti-A7075-Nb)

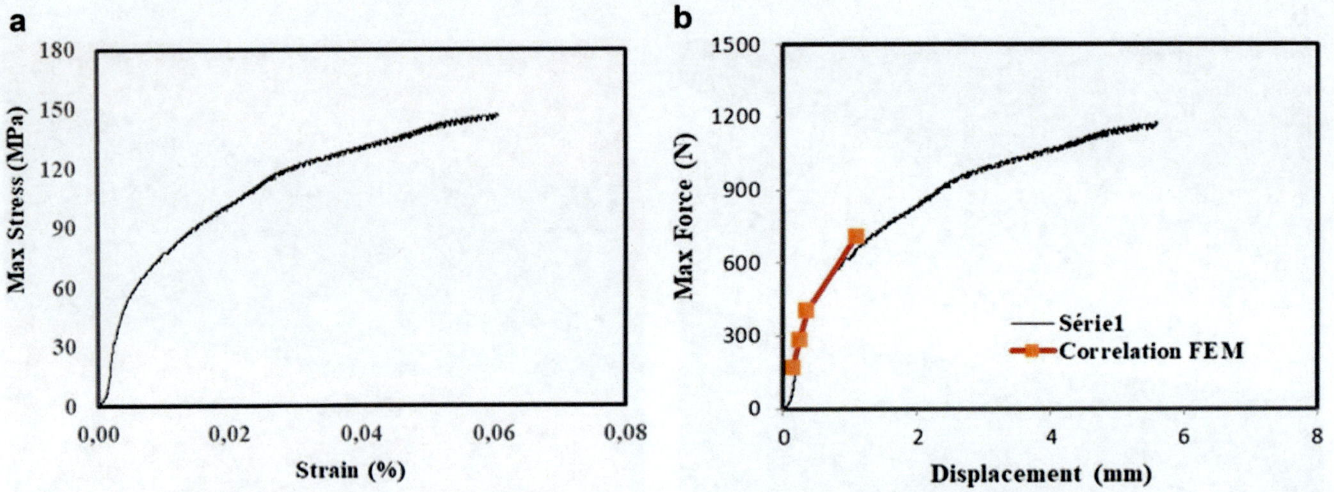

Fig. 5 (**a**) 3PB test results: stress strain values to compare with numerical solution and (**b**) presentation numerical correlation with force vs displacement evolution on the same graph

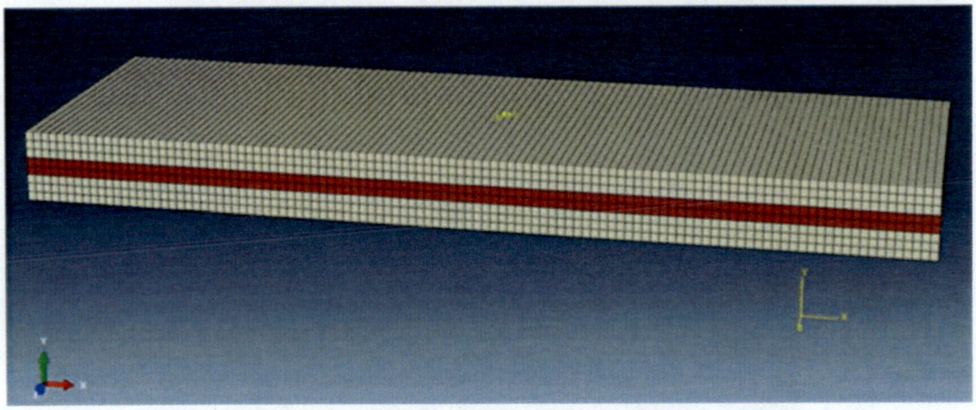

Fig. 6 Design of numerical correlation for the sandwich structure with the specimen no. 1, AA7075-Ti-Nb composite to observe maximum stress concentration on the specimen geometry

A finite element model (FEM) was created using ABAQUS software, representing the 3PB test of this specimen. The model was created by representing both the sandwich structure (in a dark colour) and the titanium (in red). The elements used were an eight-node linear hexahedral solid element with reduced integration (C3D8R). Using the reverse engineering method, the established homogenized "Young's Modulus" of the sandwich structure (dark elements in FEM) was $E = 10{,}896$ MPa. Using the stress/strain curve in the sandwich structure (Fig. 5b) the plasticity was considered to have isotropic behaviour. As will be seen in the next images, there is an significant rate of plasticity for both the A7075-Ti-Nb (A7075 and Ti) sheets.

The results of the FEM analyses are shown in Fig. 6; by this figure, a good correlation of the FEM behaviour compared with the experimental results is obtained.

An example of the maximum Von Mises stress location is shown in (Fig. 7). It seems that the maximum stress values were located only in the side of the titanium sheet near the interface with the sandwich structure. This result gave only one sandwich structure with the specimen no. 1.

As the second series, 3PB fatigue tests have been conducted on the sandwich structures to evaluate time-dependent behaviour such as stiffness and displacement amplitude.

All the tests have been carried out on the fatigue test device developed at Airbus/Isae-Supmeca-Paris (Dr G. Zambelis) with the maximum stress controlled at 8 Hz in Mode II. A prestress was applied for each specimen to evaluate stiffness and displacement amplitude during the solicitation. Stiffness values depending on the key parameters obtained from fatigue tests have been calculated according to the following equation.

Fig. 7 Correlation of the FEM behaviour with the experimental results. Evolution of Von Mises stress located only in the titanium sheet side near the interface

$$\text{Stiffeness}\left(\text{in}\ \frac{\text{N}}{\text{mm}}\right)=\frac{F\ (\text{Amplitude})}{D\ (\text{Amplitude})}=\frac{F_{\max}-\frac{F_{\max}+F_{\min}}{2}}{D_{\max}-\frac{D_{\max}+D_{\min}}{2}}.$$

Figure 8 gives the results for three different sizes (geometrical effect) of the sandwich structures (Ti-A7075-Nb). Each specimen gives a certain evolution of the stiffness and displacement amplitude depending on the number of cycles.

As shown in the figure, evolution of the maximum stress controlled is always constant to show the reliability of the fatigue tests. As observed again from these graphs, stiffness decreases in a regular way whereas the displacement amplitude increases during the test.

Fatigue life for the specimens tested in the present work varies from 1 to 6. 10^6 is the number of cycles.

As the third series, static tensile tests have been conducted on the sandwich structures to evaluate the mating effect on the mechanical behaviour of the sandwich structure composite. All the tests have been carried out on the Zwick test device in the composite laboratory of ISAE-Supmeca-Paris under quasi-static tensile test conditions with a tensile speed of 0.5 mm/min using a load cell of 100 kN.

In Fig. 9, the geometry of the test specimens, tensile test device and deformed test specimens at the level of the holes of both the upper and lower parts of the specimens. During the test, in the upper part of the specimen the hole was deformed (elliptic shape) in a regular way, but only one or two specimens were deformed at the lower hole of the specimen (not regular lower hole deformation).

Figure 10 present SEM microscopic results showing shearing failure (two cracks) at the level of the upper hole of the specimens. The mating effect of the sandwich structure should be evaluated first by measuring the shearing and then should be simulated for each condition of the test specimens. Within the framework of the common research, this type of test should be repeated and compared even with other sandwich structures containing different composites.

The preliminary results presented here justify the idea that the mating effect can cause heavy shearing on the holes of the specimens. In the case of A7075-Ti-Nb sandwich structure tests, delamination occurred at the upper holes of the tensile sandwich structure sheets [7, 9].

Figure 11 Summary of the tensile test results carried out on the sandwich structural sheets for different specimen sizes. All of the sandwich structure has shown a mating effect.

Conclusion

A new design sandwich structural composite developed A7075-Ti-Nb intermetallic composites from recycled fresh scrap-sheets supplied by a French aeronautical company. The composites were manufactured through hot forging diffusion bonding process.

In the present work, three different tests were conducted on these composites: static and cyclic, fatigue (time dependence) 3PB test and also static tensile tests on the sandwich structures containing two holes to evaluate the mating effect of these composites.

For the fatigue test, a simple and efficient stiffness and displacement amplitudes were measured depending on the number of cycles under stress-controlled conditions. In this way, a low-cost manufacturing process was developed for aeronautical engineering applications. In summary, encouraging results were obtained from the tests performed.

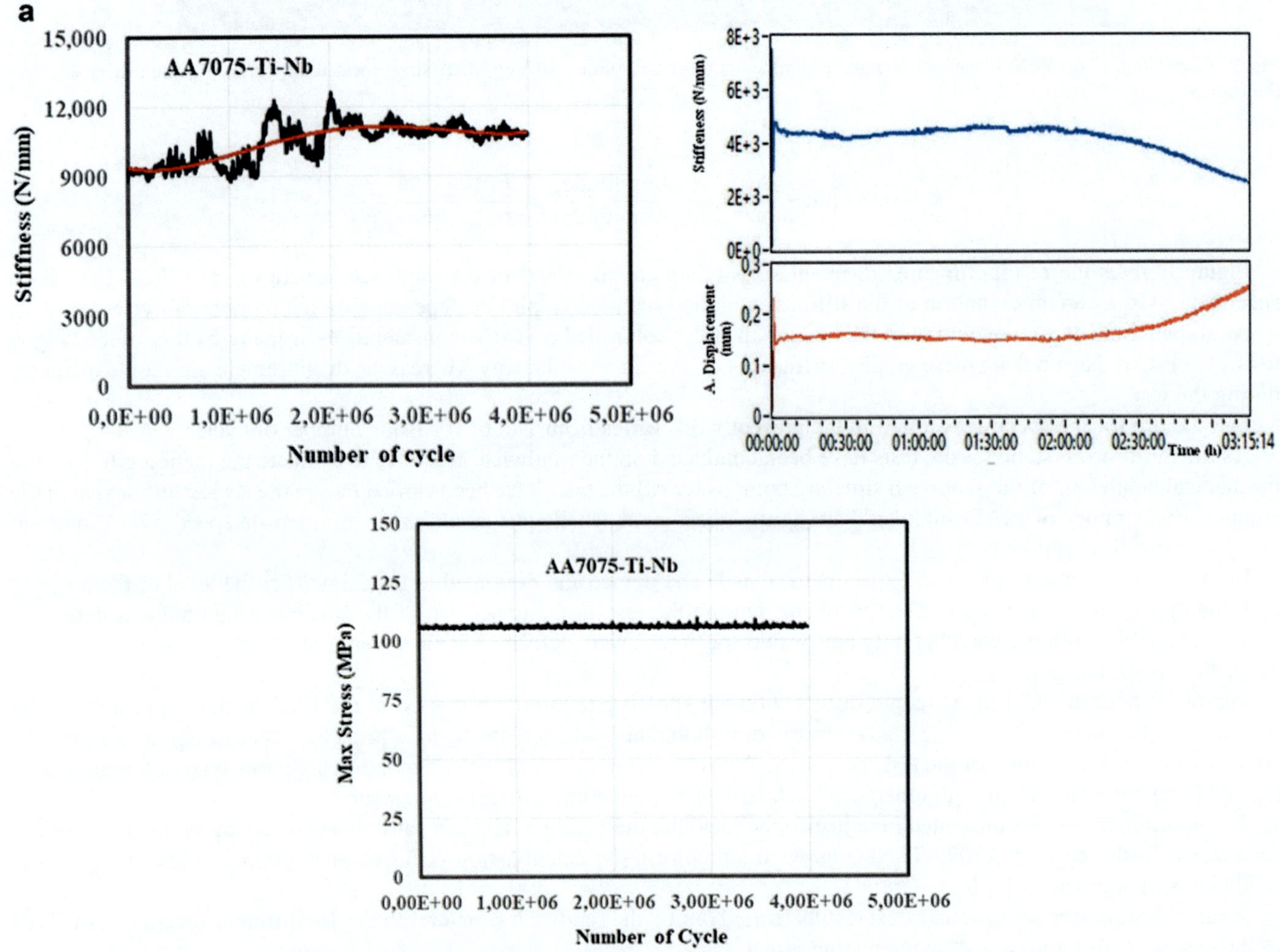

Fig. 8 (**a**) Fatigue test results for AA7075-Ti-Nb with the specimen no. 1 sandwich structure composite. Evolution of the stiffness and stress amplitude as a function of the number of test cycles. (**b**) Fatigue test results for AA7075-Ti-Nb with the specimen no. 2 sandwich structure composite. Evolution of the stiffness and stress amplitude as a function of the number of test cycles. (**c**) Fatigue test results for AA7075-Ti-Nb with the specimen no. 3 sandwich structure composite. Evolution of the stiffness and stress amplitude as a function of the number of test cycles

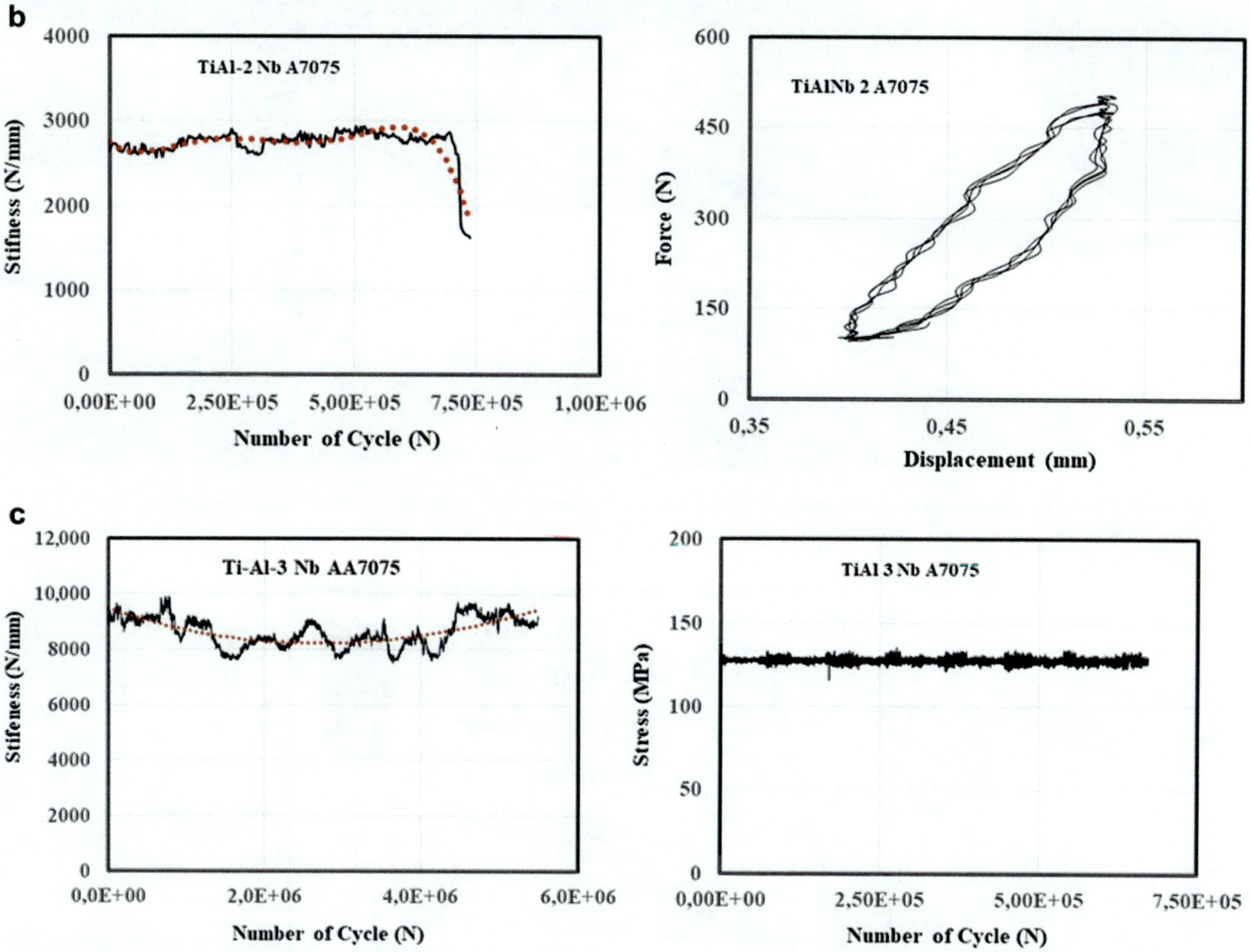

Fig. 8 (continued)

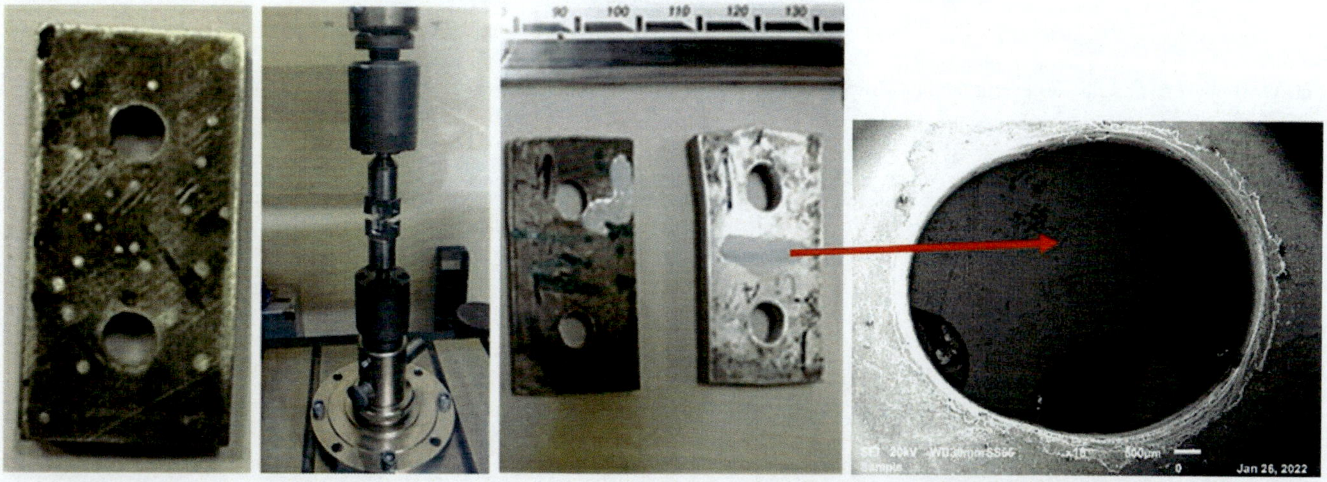

Fig. 9 Geometry of the test specimens and tensile test device (left) and deformed test specimens at the level of the holes of both the upper and lower parts of the specimens (right)

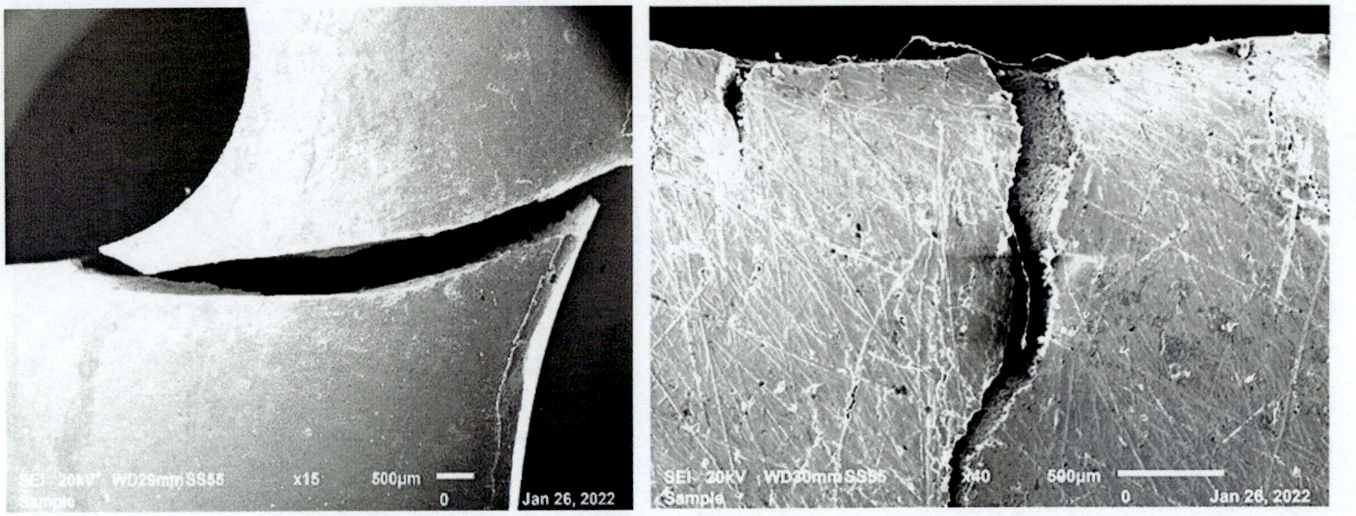

Fig. 10 Scanning Electron Microscopy results showing shearing failure at the level of the upper holes of the sandwich structural specimens

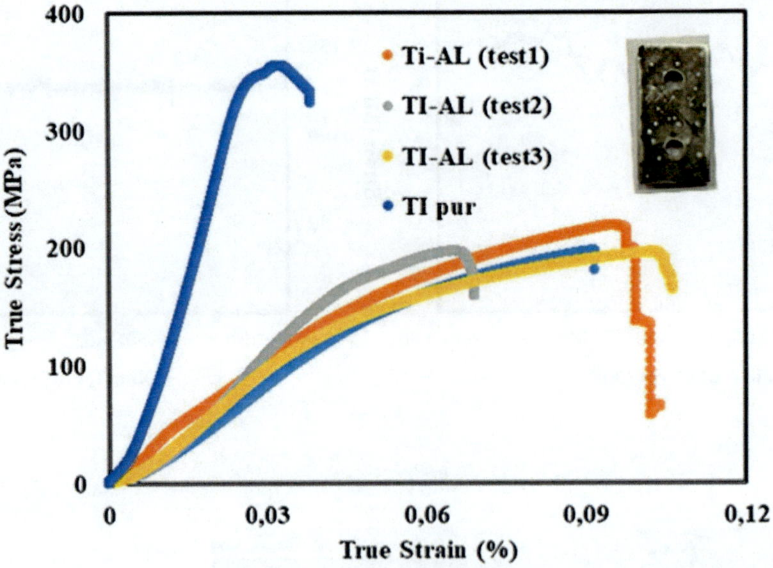

Fig. 11 Tensile test results carried out on the sandwich structural composite sheets (A7075-Ti-Nb) compared with pure titanium sheets

Acknowledgements Authors thank for the fatigue test design the assistant engineer Mr Christophe Ben Brahim and Airbus Helicopter for the static 3PB test facilities. Mechanical tests (Zwick test device) and microstructural analyses were carried out at ISAE-SUPMECA-Paris.

References

1. Munoz-Morris, M.A., Rexach, J.I., Lieblich, M.: Comparative study of Al-TiAl composites with different intermetallic volume fractions and particle sizes. Intermetallics. **13**, 141–149 (2005)
2. Peng, H., Jicai, F., Yiyu, Q.: Analysis of diffusion bond interface of TiAl base alloy with Ti, TC4 alloy and 40Cr steel. J. Harbin Inst. Technol. **7**(2), 78–81 (2000)
3. Gatamorta, F., Miskioglu, I., Katundi, D., Bayraktar, E.: Chapter 6: Recycled Ti-Al-Cu matrix composites reinforced with silicon whiskers and γ-alumina (Al2O3) fibres through sintering + forging. In: Mechanics of Composite, Hybrid & Multi-functional Materials, vol. 5, pp. 49–54. Springer, Cham (2023). https://doi.org/10.1007/978-3-031-17445-2_6

4. He, P., Feng, J.C., Zhang, B.G., Qian, Y.Y.: A new technology for diffusion bonding intermetallic TiAl to steel with composite barrier layers. Mater. Charact. **50**, 87–92 (2003)
5. Threadgill, P.L., Dance, B.G.I.: Joining of intermetallic alloys further studies. TWI J. **6**(2), 257–316 (1997)
6. Enginsoy, H.M., Bayraktar, E., Miskioglu, I., Katundi, D.: Chapter 8: Manufacturing of "Ni-Ti" based composites from fresh scrap thin sheets reinforced with Nb and TiB2 through hot-forged bonding: sandwich structures. In: Mechanics of Composite, Hybrid and Multifunctional Materials, vol. 6, pp. 53–59. Springer, Cham (2021). https://doi.org/10.1007/978-3-030-59868-6_8
7. Bayraktar, E.: Tensile test results and mating effect on the sandwich structural specimen containing different holes. Medium Internal Report, ISAE-Supmeca, Paris, 35 pages, 2022
8. Baiyun, H., Zhougyong, D., Yuehui, H.: Superplastic behaviour of a fine grained TiAl based alloy treated by multi-step thermo-mechanical treatment. Trans. Nonferrous Met. Soc. China. **8**(1), 107–108 (1998)
9. NGAINDJO Rohane Landry, PACELLI Jordan: Final report of MSc on the mating effect of the TiAl/Nb sandwich sheet structures, ISAE-Supmeca, 2022
10. Katundi, D., Miskioglu, I., Bayraktar, E.: Chapter 4: Design of intermetallic Mg (recycled Ti-Al) based composites through semi powder metallurgy method. In: Mechanics of Composite and Multi-functional Materials, vol. 5, pp. 27–33. Springer, Cham (2019). https://doi.org/10.1007/978-3-030-30028-9_4

Development of Self-Healing Glass Fiber–Reinforced Laminate Composites for Wind Turbine Blades

M. Atif Yilmaz, Kemal Hasirci, Hasan Yakar, Serhat Cetin, Deniz Isık, and Alaeddin Burak Irez

Abstract Among the various renewable energy sources, wind energy offers an effective solution to energy providers. Onshore wind turbines are generally designed for sites with low wind resources, while offshore wind turbines can be more efficient in producing energy, thanks to their longer blades that provide more than 10 MW of rated power. Offshore wind turbine blades are subjected to significantly higher stresses and harsh environmental conditions. Therefore, self-healing composites can offer cost-effective and long-lasting solutions for wind turbine blade manufacturers since self-healing is a prominent mechanism used in various industrial applications to repair the structures in the presence of a crack. In this study, the advantage of using self-healing mechanism in laminate composites is studied. Following manufacturing of self-healing microcapsules, they are incorporated into laminate composites by means of vacuum-assisted resin transfer molding (VARTM) method. Then, mechanical characterizations and microscopic examinations are carried out through tensile and Charpy impact tests. In this study, it is intended to examine the mechanical influence of using the self-healing microcapsules, as well as the curing scenario is analyzed in detail by comparing the test results. It is seen that the self-healing agent ratio of 2.5% has the optimum ratio when we compare it with 5% and 7.5% healing agent ratio. In addition to that, curing temperature of 100 °C increases the UTS when it is compared with 80 °C.

Keywords Offshore wind turbines · Laminate composites · Glass fiber · Self-healing

Introduction

Over the last decades, wind energy has become one of the main natural energies exploited by the installation of wind farms on sea and on land. The biggest challenge is to provide maximum efficiency with minimum cost. Therefore, new materials have been developed for wind turbine blades to improve their strength and stiffness [1].

Offshore turbines, in particular, are subjected to high wind speeds and environmental damage. Furthermore, the initial investment and maintenance expenses for offshore turbines are higher than those for onshore turbines. Therefore, it is crucial to develop durable turbine blades to address the above-mentioned environmental conditions and strong wind potential. Besides this, the main loads that wind turbine blades are subjected to are structural loads, which are primarily weight, wind, and centrifugal forces, and impact loads such as bird strikes [2]. Bird strikes can cause substantial damage to composite structures such as aircraft wings and wind turbine blades, as these loads produce microcracks in the structure and cause the composite blade to fracture following crack propagation [3]. In the event of a bird strike, blade maintenance may be required, and the cost of maintaining turbine blades, particularly offshore turbines, represents a significant portion of the operational cost of the blades. Improving reliability and structural endurance is therefore key to the design [4].

Microcrack growth and coalescence would result in catastrophic failure of the materials and shorten their service lifetime. In order to eliminate the hidden damage, early detection, diagnosis, and repair of microcracks become significant. Therefore, materials with a self-healing capability are appropriate for long-term operation [5].

Throughout the last few decades, self-healing materials have been developed to an important stage in self-healing research for composite structures. Self-healing composites have been created to repair cracks and damage without causing structural failure, reducing structure maintenance costs due to increased reliability. These enhancements have mostly been applied for

M. A. Yilmaz
Department of Defence Technologies, Graduate School, Istanbul Technical University (ITU), Istanbul, Turkey

K. Hasirci · H. Yakar · S. Cetin · D. Isık · A. B. Irez (✉)
Department of Mechanical Engineering, Faculty of Mechanical Engineering, Istanbul Technical University (ITU), Istanbul, Turkey
e-mail: irez@itu.edu.tr

F. Gardea et al. (eds.), *Mechanics of Composite, Hybrid and Multifunctional Materials, Volume 5*,
Conference Proceedings of the Society for Experimental Mechanics Series, https://doi.org/10.1007/978-3-031-50478-5_4

fiber-reinforced polymer composites, fiber-reinforced ceramic composites, and metal alloy composites, and have been used in a wide range of industries, particularly advanced aerospace applications [6].

For fiber-reinforced polymer composites, various self-healing technologies are currently under development. Most of these technologies have been influenced by observations of nature and are bioinspired. The most recent research on self-healing has attempted to mimic natural healing by examining the vascular networks observed in biological systems and mammalian blood clotting [7]. Research shows that three conceptual strategies have been used to demonstrate self-healing: vascular healing systems, intrinsic healing polymers, and capsule healing systems. Self-healing, without human intervention, may require external energy or pressure. All types of polymers, including thermosets, thermoplastics, and elastomers, have the ability to self-heal. To date, most studies have focused on the recovery of mechanical integrity after quasi-static fracture [8].

Materials with self-repair capability repair themselves and automatically recover their mechanical properties. More than a decade of research on this topic has resulted in self-healing composites with more than 100% efficiency in attribute recovery, including interlaminar fracture toughness [9]. For example, in double cantilever beam fracture specimens with tapered widths where delamination in the medial plane is introduced and then allowed to heal, self-healing is observed. When autonomous self-healing occurs at room temperature, up to 45% of the original interlaminar fracture strength can be recovered; however, when healing occurs at 80 °C, the recovery rate improves to over 80% [10]. In another study, a new self-healing technique that uses a solid-state thermoset repair system was shown to recover 50–70% of its failure strength after healing. E-glass composites made with this resin demonstrate that it is possible to reduce the delamination zone and repair matrix cracks [11]. The concentration of the solid-state healing agent in a standard epoxy resin is also optimized and the effectiveness of the modified resin in healing damage in composites is examined. Composites manufactured from a resin containing 7.5 wt% of the healing agent are found to have optimal healing ability in Charpy test specimens [12].

This research presents the self-healing of composite structures for wind turbine blade applications. Literature survey given above shows that many studies are observed in self-healing for composite structures; however, to the best of authors' knowledge, no study examined the effect of healing agent weight ratio and curing temperature on the mechanical properties of these composites with different control groups. Then, this study aims to fill this gap with the following research. Damage was created on the specimens by Charpy impact test. Then, the specimens were cured at different temperatures. Finally, the mechanical properties of the specimens were examined by tensile tests.

Materials and Methods

Epoxy-based composites are manufactured with unidirectional glass fiber fabrics in this study. L300 unidirectional E-glass fiber fabrics were used for manufacturing laminated composites. The laminate composites consist of eight layers of fabrics as [0/45/90/-45]s (see Fig. 1).

Fiber and matrix were used with a weight ratio of 1:1. Hexion LR285 epoxy resin and Hexion LH285 hardener were used with a weight ratio of 5:2 (see Table 1).

Composites were manufactured by means of the vacuum-assisted resin transfer molding (VARTM) method. Glass fibers were laid by hand according to the predetermined angles as given in Fig. 1, epoxy resin was impregnated to the glass fibers with the assistance of vacuum pressure, and healing agents are spread between each layer equally. Composites are manufactured as plates in order to get specimens from these plates. Hexion LR285 epoxy and LH287 hardener were used with the ratio of 100:40, respectively.

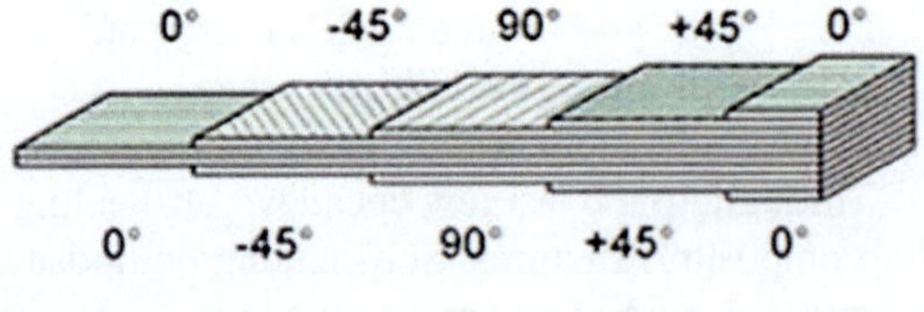

Fig. 1 Stacking sequence of unidirectional glass fiber fabrics

Table 1 Density and viscosity values for epoxy and hardener

Material	Density (g/cm^3)	Viscosity (mPas)
Hexion LR285 epoxy	1.18–1.23	600–900
Hexion LH287 Hardener	0.94–0.97	80–120

Table 2 Test scenario for composites with/without self-healing

Self-healing agent/epoxy ratio (%)	Self-healing agent	Curing conditions	Specimen #
0	No self-healing	–	Specimen 1.1
0	No self-healing	Damaged	Specimen 1.2
2.5	With self-healing agent	16 h 25 °C + 8 h 80 °C	Specimen 2.1
5	With self-healing agent	16 h 25 °C + 8 h 80 °C	Specimen 3.1
5	With self-healing agent	16 h 25 °C + 8 h 100 °C	Specimen 3.2
7.5	With self-healing agent	16 h 25 °C + 8 h 80 °C	Specimen 4.1

Fig. 2 VARTM manufacturing setup for glass fiber composite plate with 7.5% healing agent ratio

The manufacturing of the composites is planned as given in Table 2. The plates are manufactured with different healing agent/epoxy ratio as 0%, 2.5%, 5%, 7.5% to investigate the effect of the self-healing material on mechanical properties as shown in Fig. 2. On the other hand, different curing temperatures are implemented on control groups for 5% healing agent ratio.

Composite structures were manufactured as composite plates, and the specimens were prepared from the composite plates by using abrasive water jet so that the mechanical properties of the specimens could be determined by tensile and Charpy impact tests. ASTM D3039 standard has been used for tensile tests in order to determine the fundamental mechanical properties.

Six rectangular specimens of 250 mm × 25 mm were prepared from the manufactured composites as specified in Table 2. Charpy impact tests were performed using a universal Charpy tester according to ASTM 6110 [13] standard to create microcrack in the composites to investigate impact load like bird strike. Therefore, a bird strike scenario is assumed for wind turbine blade and a CFD analysis has been executed to calculate the speed of the bird for 1.8 kg (4 lbs.) according to EASA Standard [14]. The results of the CFD lead to bird strike speed of 11.4 m/s where kinetic energy can be calculated as:

$$\text{Energy} = \frac{1}{2}mV^2 = \frac{1}{2}(1.8)(11.4)^2 = 116.09\ \text{J}$$

The effective area for impact was calculated in one of the studies as 750 × 500 = 375,000 mm^2 [15]. However, the effective area for Charpy impact test is 185 × 25 = 4625 mm^2 in our research although the test specimen has 250 mm × 25 mm dimensions. Since the effective area for our study is 1/80 of the reference study, the energy for the Charpy impact test is calculated as:

$$\text{Charpy energy} = 116.09/80 \cong 1.5\ \text{J}$$

After the impact test, tensile tests were carried out by using Shimadzu AG-x 50 kN testing machine with a loading rate of 2 mm/min. Elastic modulus, ultimate tensile strength, and strain at break were determined via tensile tests according to ASTM D3039 [16].

Table 3 Tensile test results of self-healing glass fiber laminated composite

Specimen #	wt% self-healing	Curing conditions	Elastic modulus (GPa)	UTS (MPa)	Strain at break
1.1	0	–	15.232 ± 1.25	337.43 ± 6.15	0.022 ± 0.01
1.2	0	Damaged	14.897 ± 3.45	328.23 ± 5.78	0.022 ± 0.01
2.1	2.5	16 h 25 °C + 8 h 80 °C	13.612 ± 7.25	334.31 ± 8.18	0.024 ± 0.02
3.1	5	16 h 25 °C + 8 h 80 °C	13.222 ± 2.15	301.55 ± 9.27	0.022 ± 0.01
3.2	5	16 h 25 °C + 8 h 100 °C	12.559 ± 1.19	312.17 ± 4.05	0.024 ± 0.02
4.1	7.5	16 h 25 °C + 8 h 80 °C	12.537 ± 3.89	304.25 ± 3.88	0.024 ± 0.02

Results and Discussion

The research was carried out to see the effect of the self-healing agent ratio and curing temperature on the mechanical properties of the composite structures. Since the dimensions of the specimen and the impact on the specimen (1.5 J) are small, we can expect microcracks in the structure. These microcracks explain 9 MPa (2.6%) decrease on UTS when we compare Specimen #1.1 and Specimen #1.2 according to the tensile test results given in Table 3. Additionally, Specimen 2.1 has the UTS of 334 MPa, which is similar to the UTS of Specimen 1.2. By comparing specimen 2.1 with specimen 3.1, the self-healing agent is increased twice; however, the UTS is decreased. It can be considered that increasing the ratio of self-healing agent from 2.5 to 5 has no significant effect on the structure. On the other hand, by comparing specimen 3.1 and specimen 3.2, one can see that increasing the curing temperature has a positive effect on the fracture strength.

According to Table 3, when the self-healing ratio increased to 7.5%, we can see that there is no significant impact on the UTS values. Therefore, it is understood that 2.5% self-healing ratio gives the optimum healing agent ratio, and increasing the cure temperature from 80 to 100 °C for 8 h also increases the UTS of the composite structure. The results shown in Table 3 point that the increased curing temperature helps healing agents to have better mechanical properties within the same time interval.

Conclusion

In this research, self-healing glass fiber–reinforced laminate composites for wind turbines were developed. VARTM method has been used for manufacturing of the composite plates, and the specimens have been obtained from these plates. Charpy impact tests were performed using a universal Charpy tester in order to create predefined damage on the structure. Then, the specimens were cured according to the effect of curing conditions in addition to the effect of self-healing/epoxy ratio. The mechanical properties were determined by tensile tests on the specimens.

As explained in the analysis section, it is seen that 2.5% healing agent ratio is the optimum ratio with the curing temperature of 100 °C. We can expect higher UTS values if we have curing temperature of 100 °C for 2.5% healing agent ratio.

By considering the results of the research, self-healing composite structures can be used for wind turbine blades according to the damage risk that the structure can continue operation. Curing conditions and self-healing ratio can be optimized for different load cases and maintenance costs can be reduced, especially for offshore wind turbines. For future studies, hybrid composite structures can be combined with self-healing capability to examine the improvement in mechanical properties of the composite structures.

Acknowledgments This research was supported by the Istanbul Technical University Office of Scientific Research Projects (ITU BAPSIS), under grant MGA-2022-43400.

References

1. Reddy, S.S.P., et al.: Use of composite materials and hybrid composites in wind turbine blades. Mater. Today Proc. **46**, 2827–2830 (2021)
2. Hüppop, O., Dierschke, J., Exo, K.M., Fredrich, E., Hill, R.: Bird migration studies and potential collision risk with offshore wind turbines. Ibis. **148**, 90–109 (2006). https://doi.org/10.1111/j.1474-919X.2006.00536.x
3. Long, S., et al.: Failure modeling of composite wing leading edge under bird strike. Compos. Struct. **255**, 113005 (2021)
4. Kanu, N.J., et al.: Self-healing composites: a state-of-the-art review. Compos. A: Appl. Sci. Manuf. **121**, 474–486 (2019)
5. Yuan, Y.C., et al.: Self healing in polymers and polymer composites. Concepts, realization and outlook: a review. Express Polym Lett. **2**(4), 238–250 (2008)
6. Das, R., Melchior, C., Karumbaiah, K.M.: Self-healing composites for aerospace applications. In: Advanced Composite Materials for Aerospace Engineering, pp. 333–364. Woodhead Publishing, Duxford (2016)
7. Trask, R.S., Williams, H.R., Bond, I.P.: Self-healing polymer composites: mimicking nature to enhance performance. Bioinspir. Biomim. **2**(1), P1 (2007)
8. Blaiszik, B.J., et al.: Self-healing polymers and composites. Annu. Rev. Mater. Res. **40**, 179–211 (2010)
9. Wang, Y., Pham, D.T., Ji, C.: Self-healing composites: a review. Cogent Eng. **2**(1), 1075686 (2015)
10. Kessler, M.R., Sottos, N.R., White, S.R.: Self-healing structural composite materials. Compos. A: Appl. Sci. Manuf. **34**(8), 743–753 (2003)
11. Hayes, S.A., et al.: A self-healing thermosetting composite material. Compos. A: Appl. Sci. Manuf. **38**(4), 1116–1120 (2007)
12. Hayes, S.A., et al.: Self-healing of damage in fibre-reinforced polymer-matrix composites. J. R. Soc. Interface. **4**(13), 381–387 (2007)
13. American Society for Testing and Materials: Standard Test Method for Determining the Charpy Impact Resistance of Notched Specimens of Plastics (ASTM Standard No. D6110-18). ASTM, West Conshohocken (2018) https://www.astm.org/Standards/D6110
14. European Aviation Safety Agency: Compliance with CS-25 Bird Strike Requirements (EASA CM – S – 001 Issue: 01). EASA (2012) https://www.easa.europa.eu/sites/default/files/dfu/certification-docs-certification-memorandum-'final'-EASA-CM-S-001-Issue-01_Compliance-with-CS-25-bird-strike-requirements_PUBL.pdf
15. Iannucci, L., Donadon, M.: Bird strike modeling using a new woven glass failure model. In: 9th International LS-DYNA Users Conference, vol. 575 (2006)
16. American Society for Testing and Materials: Standard Test Method for Tensile Properties of Polymer Matrix Composite Materials (ASTM Standard No. D3039/D3039M-17). ASTM, West Conshohocken (2017) https://www.astm.org/Standards/D3039

Development of Recycled Aluminium (AA7075 + AA1050)-Based Hybrid Composites Reinforced with Recycled Rice Husk Produced by Sintering + Forging

Fabio Gatamorta, Noé Cheung, Dhurata Katundi, Ibrahim Miskioglu, and Emin Bayraktar

Abstract A new recycled hybrid composite has been designed by using a special doping process and a combined method, "sintering + forging" of recycled "AA 7075 + AA1050" and basically used rice husk as a fine powder and graphene nanoplatelets. Static and cyclic behaviours of these composites and also time-dependent behaviour called modified fatigue behaviours have been evaluated under compression solicitation. A detailed damage analysis has been performed using scanning electron microscopy.

Keywords Recycled aluminium · Rice husk · Ceramics · SEM · Static and dynamic

Introduction

The improvement of aluminium-based hybrid composites reinforced with agro-based and or sole waste is now a new idea for environmental protection and economic composite design. These composites are now extensively found in manufacturing engineering owing to light and high strength properties. Recently, rice husk and coconut fibres etc. have been used extensively in engineering applications. As many other forms of agro-based waste are generally used in energy applications (fired, etc.), the rice husk were taken from the energy industry and used in this work as a low-cost composite design [1–7].

These hybrid composites are mainly designed with additional reinforcement to obtain great strength and wear resistance and for other reasons such as corrosion. In this work, we especially used only very fine rice husk and a small quantity of graphene nanoplatelets (GNPs) to decrease the cost of the process or to find another solution to increase the strengthening of the hybrid composites that we have worked for an aeronautic applications for secondary parts of the turbo-compressor by using AA7075 fresh scrap given by Embrayer Brazilian Aeronautical Company. As a matrix of the specimens, fresh scraps of aluminium AA7075 and A1050 are recycled for a homogenous mixture before the addition of the reinforcement. Rice husk and GNPs were added into the matrix by a special doping process at 250 °C.

Because GNPs have more advantages in mechanical strength and very high resistance against wear, micron-sized reinforcements as known that GNPs are very often used in the manufacturing of different pieces in the design of new composites [6–13].

For this reason, we have used the double effect of the large amount of very fine reinforcements together with nanoparticles in small quantities. All of the manufacturing has been made as a solid-state process.

F. Gatamorta (✉)
State University of Rio de Janeiro, Rio de Janeiro, Brazil
e-mail: fabio.gatamorta@uerj.br

N. Cheung
UNICAMP, University of Campinas, FEM, Campinas, SP, Brazil

D. Katundi · E. Bayraktar
ISAE-Supmeca-Paris, School of Mechanical and Manufacturing Engineering, Paris, France
e-mail: emin.bayraktar@isae-supmeca.fr

I. Miskioglu
Michigan Technological University, ME-EM Department, Houghton, MI, USA

F. Gardea et al. (eds.), *Mechanics of Composite, Hybrid and Multifunctional Materials, Volume 5*,
Conference Proceedings of the Society for Experimental Mechanics Series, https://doi.org/10.1007/978-3-031-50478-5_5

Our research groups UNICAMP-FEM, ISAE-SUPMECA, MICHIGAN TECH USA have developed an innovative process to improve the mechanical properties and increase the toughening mechanisms of rice husk/GNPs micro and nano reinforced recycled constituents aluminium matrix composites [1, 2, 7–11]. This paper concentrates on the toughening mechanism conducted on the static-cyclic mechanical properties, microstructural characterization, generally speaking, the strengthening mechanisms of the new hybrid aluminium-based composites.

Experimental Conditions

As a matrix for these composites, two recycled aluminium powders were used; "A7075" from the recycled chips supplied by a Brazilian aeronautic company, and recycled pure Al, A1050. First, both of the recycled aluminium chips were gas atomized and then they were mixed (50 wt%). They were prepared by high-energy milling in a planetary ball mill under an inert argon atmosphere to prevent oxidation of the powders (20/1 ball/powder ratio), at the University of São Carlos-SP-Brazil. During the second stage, 4 wt% of zinc stearate was added to the mixture for homogenous milling as a lubricant during the preparation of the composite. After the milling operation, the thermal behaviour of AA7075 alloy powder was evaluated using differential scanning calorimetry thermogravimetric analysis (DSC-TGA) and X-ray diffraction (XRD) diagrams. In fact, all of the details of the experimental conditions were given in previously published papers [2, 6, 7].

This matrix was divided into three parts and reinforced with rice husk as a major and as three minor reinforcements, molybdenum and copper (Mo 1 wt%, Cu wt1% and GNPs 0.75 wt%) were used respectively. During the milling, pure nano AA1050 (5 wt%) was added to facilitate and homogenize the mixture of the two types of recycled aluminium alloys. Rice husk was doped with matrix at 250 °C before sintering. After compaction of the homogenous mixture, sintering was carried out under argon gas in the oven. Sinter heating was conducted up to the maximum sintering temperature (600 °C) for 45 min. During the last stage, our research was conducted on the hot forging that was carried out to complete the manufacturing processes at the lower temperature than the sintering temperature. Mechanical and physical properties were evaluated through micro-hardness and detailed static-cyclic compression tests.

Results and Discussion

Figure 1 shows a DSC diagram measured for A7075 alloy and also simulation of the fraction of the solid depending on the temperature calculated using the software "Thermo-Calc" in the matrix to determine the critical transformation points during the heating and cooling stages. Additional information was given for energy-dispersive X-ray spectroscopy (EDS) chemical analysis in the next sessions for each composite.

For the rice husk, the same thermal analyses were carried out; DSC-TGA and XRD diagrams and also powder size were measured using LASER granulometry as shown in Fig. 2.

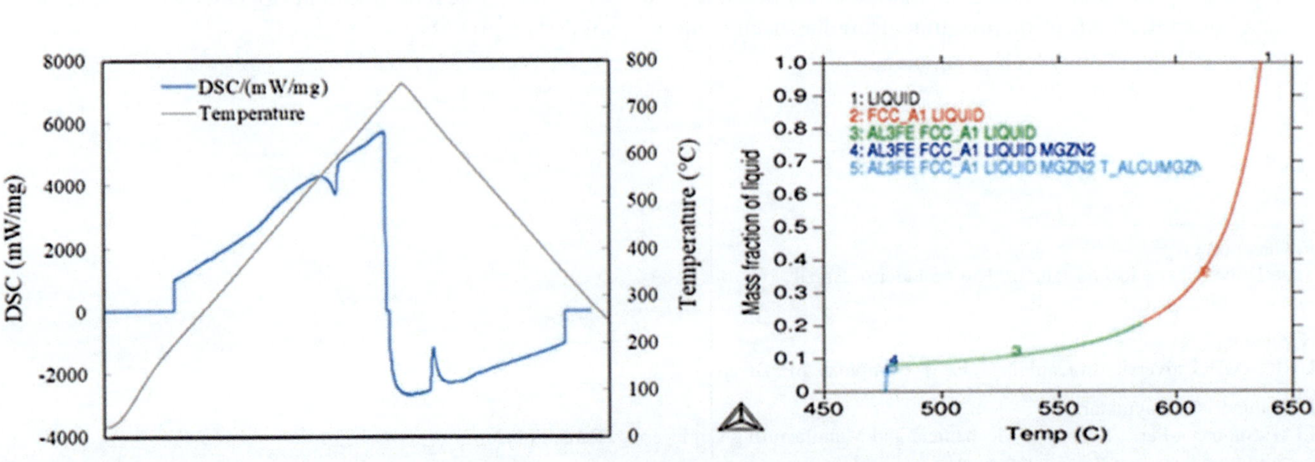

Fig. 1 Differential scanning calorimetry (DSC) diagram measured for A7075 with a heating rate of 5 °C/min and also simulation of the fraction of solid depending on the temperature, [2, 7, 10]

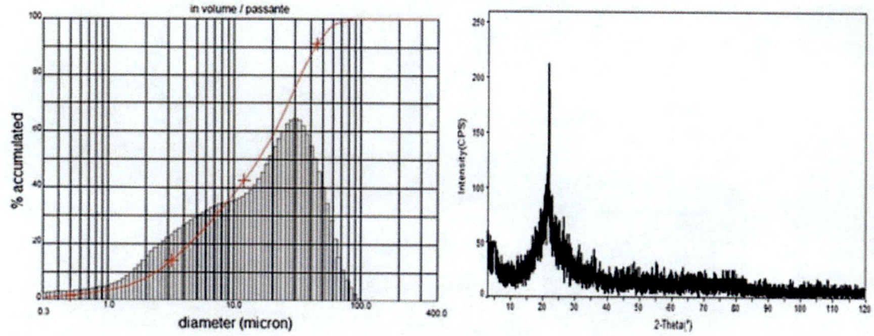

ANALYSE			
Elements	100% m/m	Oxides	100% m/m
Si	93,932	SiO_2	97,564
Cl	0,356	K_2O	1,395
K	3,541	CaO	0,586
Ca	1,325	MnO	0,336
Mn	0,845	Cl	0,119

Fig. 2 X-ray diffraction (XRD) analyses of the rice husk to detect the phases after sintering of the composite

Table 1 (a) Compositions of the A7075 alloy and (b) composition of three composites formulated for the innovative hybrid composite design together with two recycled A1050 and A7075 respectively

(a) Chemical composition of scrap AA 7075 alloy (wt%)

Element	Al	Cu	Fe	Mg	Mn	Si	Ni	Zn	Cr	Zr
wt%	Balance	1.48	0.23	2.11	0.07	0.10	0.01	5.29	0.22	0.02

(b) Compositions of the three composites (wt%)

Composite name	A1050 (50%) AA7075 (50%)	Rice husk	Cu	Mo	GNPs
N-I	B	10	1	1	0.75
N-II	B	20	1	1	0.75
N-III	B	30	1	1	0.75

Table 2 Electrical and thermal properties measured for four composites with microhardness values

Composite name	Electrical conductivity at ambient (S/m)	Thermal conductivity (W/mK)	Microhardness ($HV_{0.1}$)
N-I	5.65×10^8	4.245	125 ± 10
N-II	5.40×10^8	3.130	170 ± 20
N-III	4.35×10^8	3.110	195 ± 15

The aim of these analyses is to prepare the design of the composites correctly and determine the phases of the sintered composites.

Table 1a presents the Compositions of the A7075 alloy and Table 1b shows composition of three composites formulated for the innovative hybrid composite design together with two recycled A1050 and A7075 respectively.

Besides the major reinforcement (rice husk) and small amounts of Mo – Cu – GNPs that increase strengthening of the composites, generating a strong cohesion of the reinforcements with the matrix. For the fine distribution of the reinforcements with a fine grain size of Cu and nano Mo (<1 μm) and GNPs (500 m^2/g with surface particles) were added into the matrix for each composite. Additionally, the presence of Mo and GNPs in the structure greatly increase the mechanical resistance for toughening, mainly because of a strong cohesion by chemical bonding diffusion at the interface between the matrix and the reinforcements. We know that fine copper particles added in the composition develops the chemical bonding diffusion in the matrix well. The formation of the chemical bonding diffusion mechanism is presented in the next session, but are only indicative and should be improved with new measurements during the period of the research project.

Electrical conductivity levels were measured using an "Agilent 4338B Milli-ohm Meter" (ISAE-SUPMECA-Paris). Three specimens were taken for the measurement of each composite and then the mean values were presented in Table 2. For the measurements, the DC-regulated power supply voltage and current were set as 20 V and 20 A respectively. Data acquisition Card "NI9234" was connected in parallel with the output of the power to acquire the voltage data (voltage input accuracy was 24 bits). A high precision multi-meter "Agilent U1253N" was connected in a series to measure the current intensity (A). For the same specimens, thermal conductivity measurements were also carried out in our laboratory (ISAE-SUPMECA-Paris) and microhardness measurements taken from the samples produced under the same conditions for these composites were also presented in Table 2. All of these data for the electrical and thermal measurements were revealed using the LabVIEW program. These results should obviously be assumed to be indicative data under laboratory conditions.

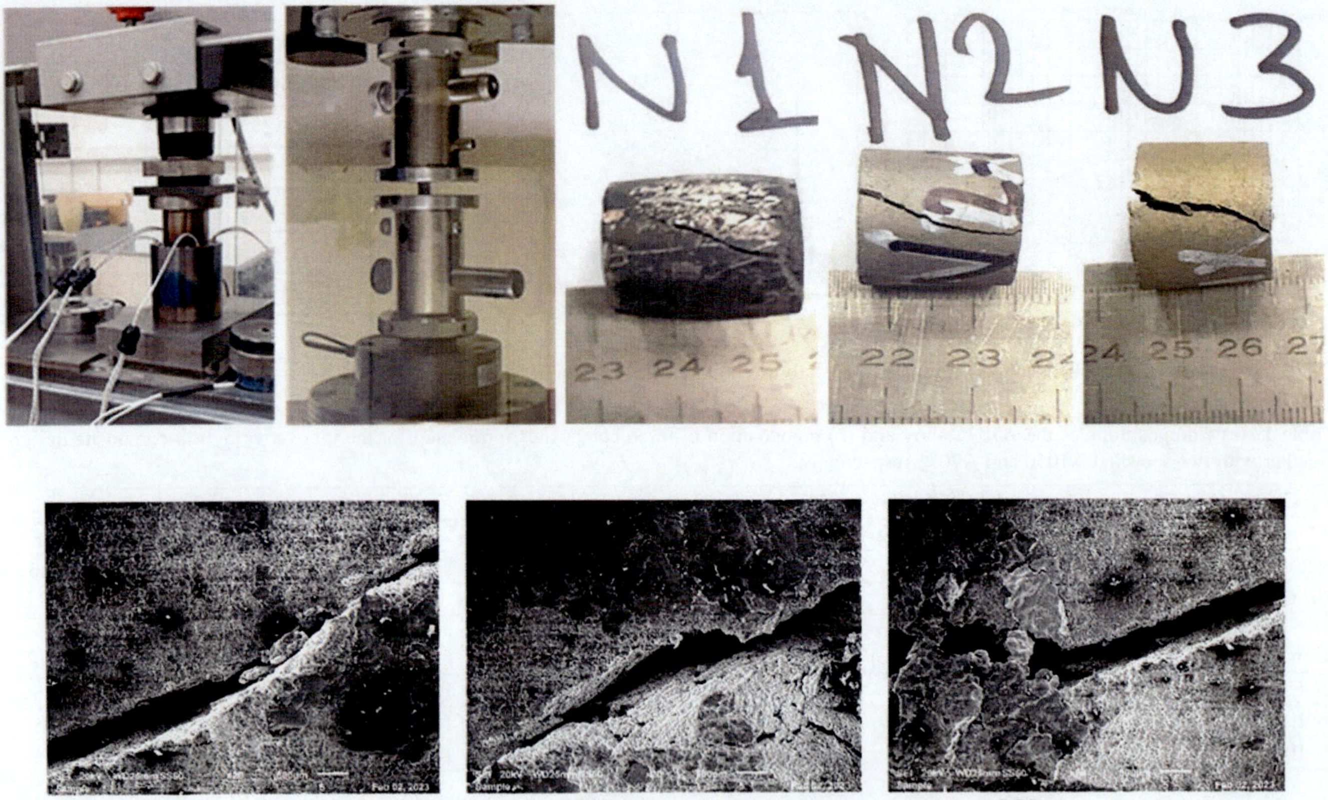

N1. N2. N3

Fig. 3 Sintering process carried out at 600 °C followed by hot forging of the specimens (left) and a compression test device adapted on the Zwick test machine (ISAE-SUPMECA/Paris), images of three failed specimens for each composition showing shear crack propagation taken SEM

In Fig. 3, a sintering process carried out at 600 °C followed by hot forging of the specimens was presented and a compression test device adapted on the Zwick test machine (ISAE-SUPMECA/Paris) was shown, with pictures of the three failed specimens for each composition showing shear crack propagation (maximum shear stress with 45° cracks) after a compression test taken from SEM.

During the second stage of this work, a detailed metallographic analysis was carried out by mapping and EDS chemical analyses for the three composites designed in the present work. Figure 4 presents this evaluation for each composition.

As regards the surface microstructures and mapping of these composites, it seems that there is homogenous distribution of the reinforced particles in the matrix.

The main reinforcement, rice husk, is given a fine homogenous distribution in the matrix; this case provides a strong cohesion with the matrix in the microstructure, which contributes well to the toughening mechanism. Some of the minor reinforcements have given local agglomeration (mainly, Mo and GNPs).

It seems that the strengthening values should be higher if the minor reinforcements could be dispersed homogenously. Within the framework of this research, these values were obtained under laboratory conditions, which make a useful tool for industrial applications.

To understand the toughening mechanism in detail, three different comprehensive tests were conducted in the present work. Static and incremental cyclic compression test results were performed to evaluate the mechanical behaviours of these innovative hybrid composites with a deformation rate of 1 and 0.5 mm/min on the Zwick test machine (ISAE-SUPMECA/Paris) with a load cell capacity of 100 kN.

Low-cycle fatigue (LCF) tests were also performed on a closed loop test device with a frequency of 0.5 Hz.

The LCF tests were performed using a computer-controlled adapter with a 100-kN load cell (ASTM E606). These tests were performed under fully reversed compression-compression loading. LCF tests were limited to 10^4 cycles in this work and maximum-applied fatigue strength was chosen around the yield strength, between 70% and 80% of the maximum strength. Figure 5 shows the total results of mechanical behaviours of the three composites under the static incremental cyclic and time-dependent LCF conditions respectively.

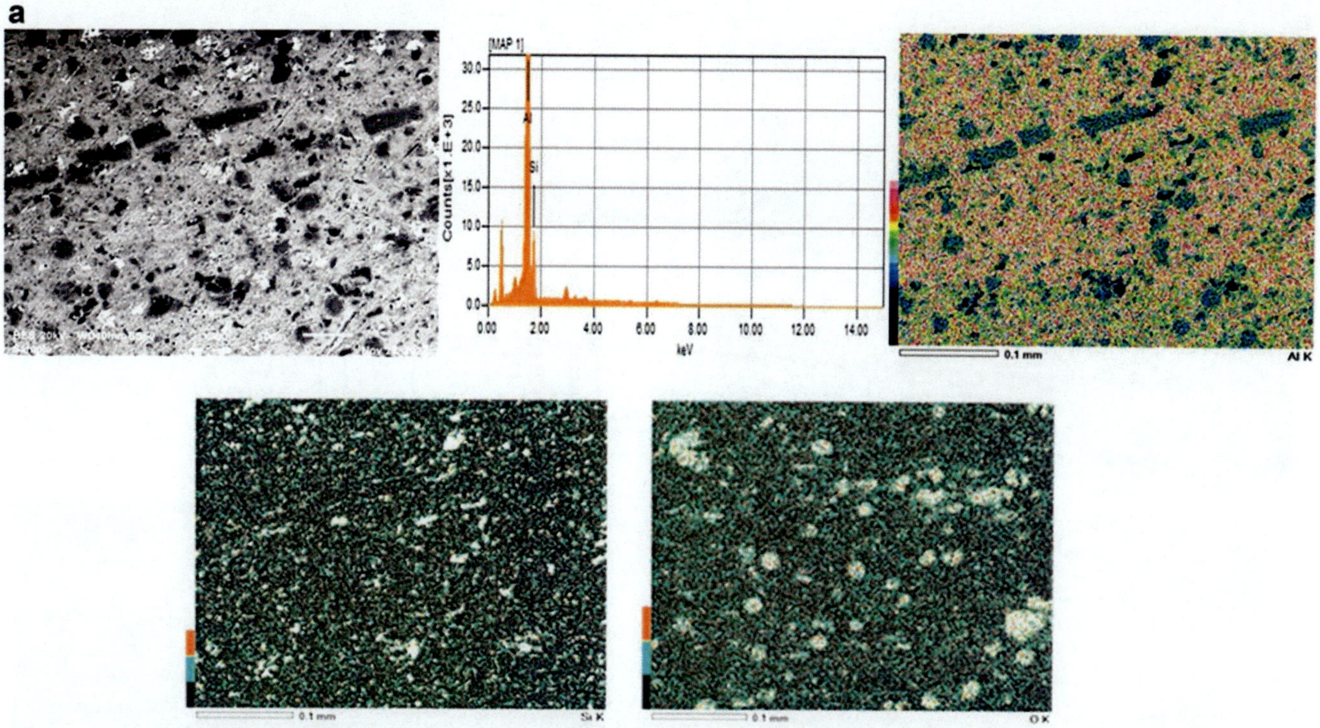

Fig. 4 (**a**) Comparative study for the microstructure of the composite N-I; mapping and energy-dispersive spectroscopy (EDS) chemical analyses (600 °C only sintering). (**b**) Comparative study for the microstructure of the composite N-II; mapping and EDS chemical analyses (600 °C only sintering). (**c**) Comparative study for the microstructure of the composite N-III; mapping and EDS chemical analyses (600 °C only sintering)

As for the fracture surfaces of these hybrid composites, most of them have given a mixed fracture behaviour, but in general, they are ductile. In Fig. 6, fracture surfaces taken from different specimens of the second composition, N-II, was presented here only as an example of these composites.

For comparison of the static compression test results, two manufacturing processes were compared here; sintering and the sintering + forging process. The second process (sinter + forging) evidently gives higher strengthening values regarding only the sintering process. Owing to the limited load cell capacity of the testing machine (Zwick test machine, ISAE-SUPMECA/ Paris), specimens produced using the sintering + forging process could not be failed after testing of the 4 specimens. Their real values are higher than the results presented here in Table 3.

All of the fatigue specimens produced from the three composites have shown ductile transgranular fatigue damage behaviour, most of them in the matrix, and having relatively large grains. The cohesion of the reinforcements with the matrix is very strong and distribution was carried out in a homogenous way each time. Agro-based waste recycled reinforcements show ductile behaviour according to the first results obtained within the framework of the composites contrary to what is indicated in the literature, that the integration of ceramic particle reinforcements into the metal matrix can decrease the ductility and toughness levels of the metal matrix composites. As we have indicated in the former section, these results are only limited to those performed under laboratory conditions.

Table 4 presents the total fatigue test results. All of the results are given here as the mean values.

It is summarized here that the test results carried out on the three composites show ductile behaviour that is related to the tough and sound microstructure with strong cohesion of the rice husk reinforcement with the matrix.

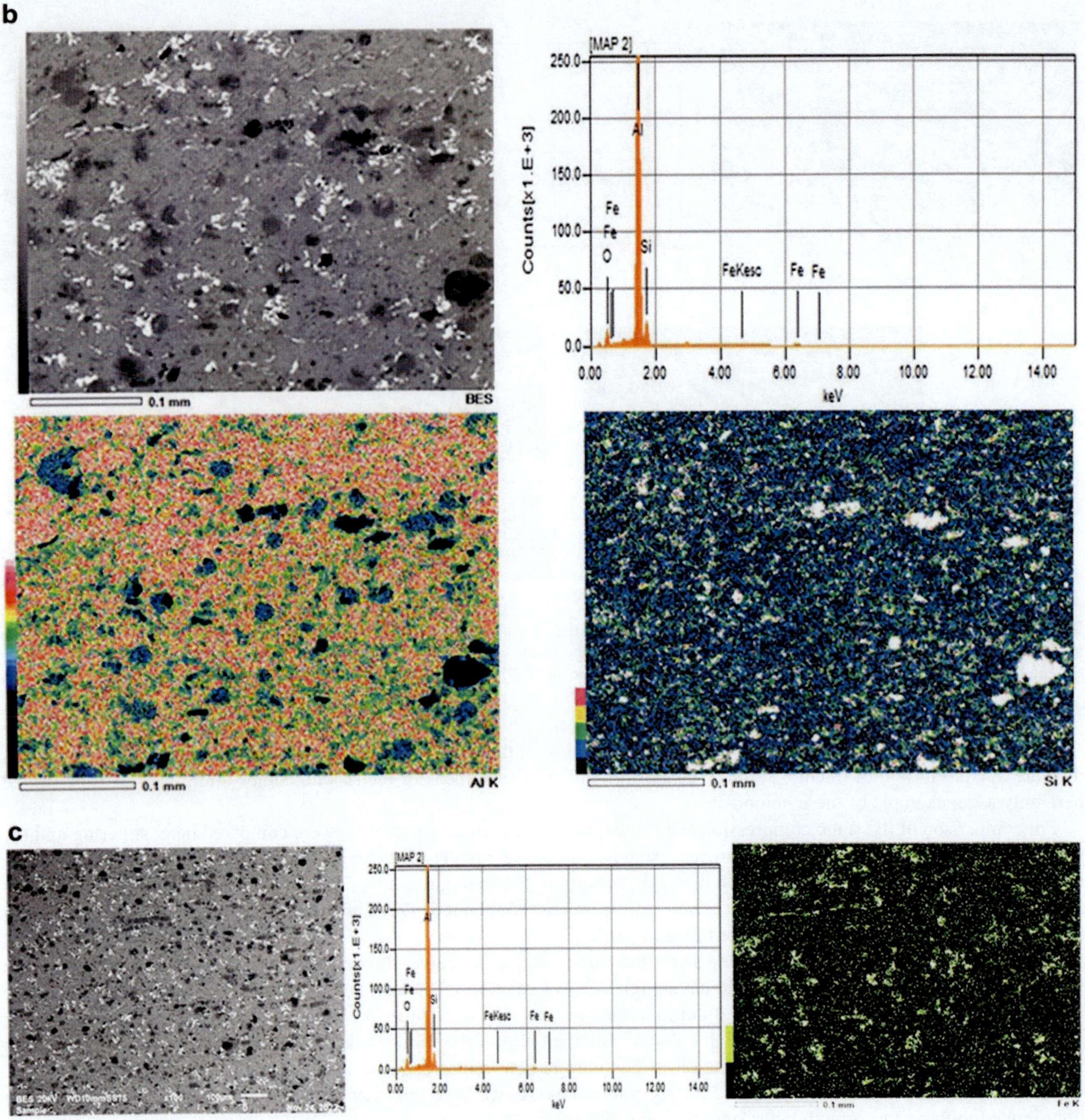

Fig. 4 (continued)

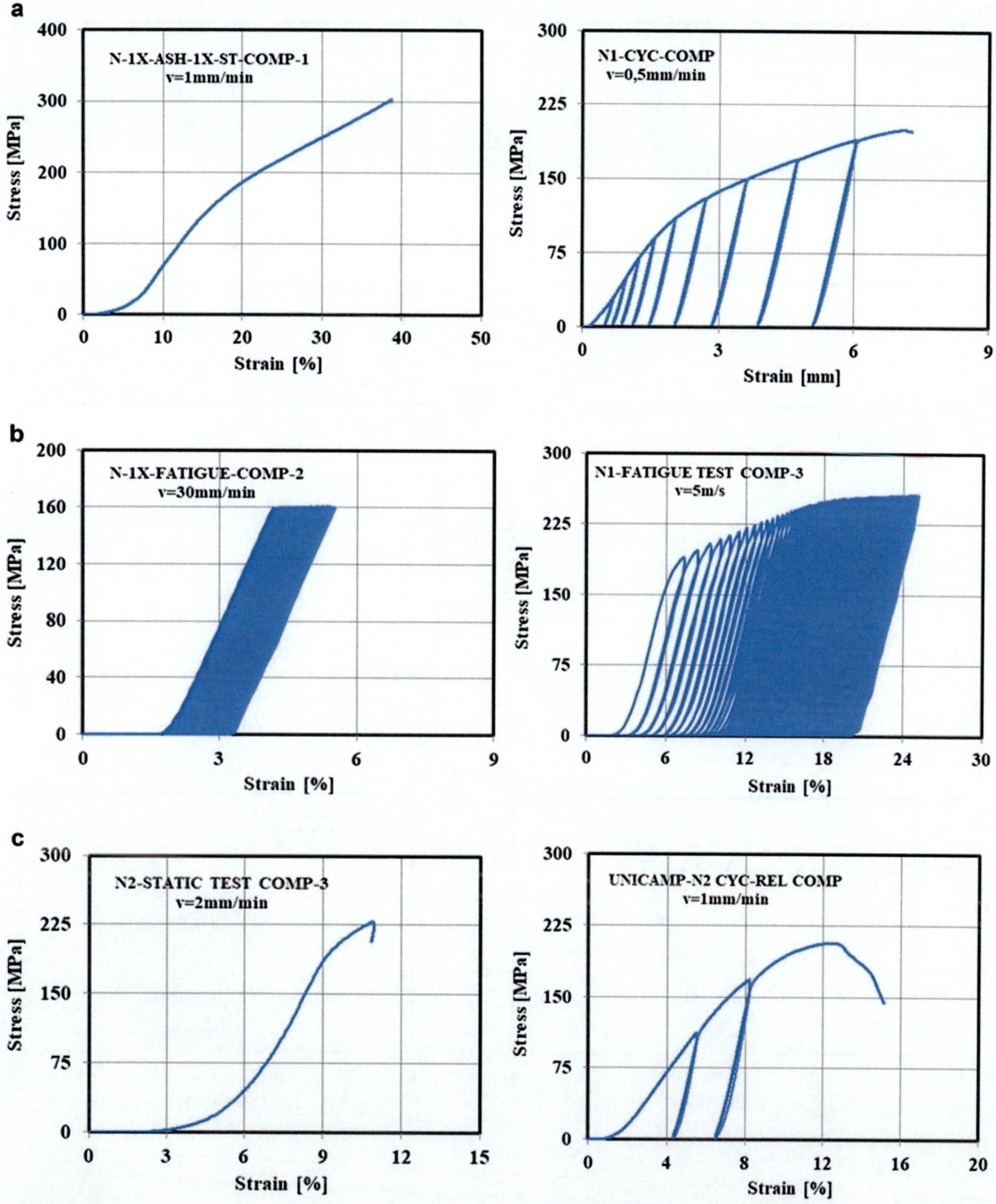

Fig. 5 (**a**) Static compression test and incremental test results carried out for the specimens taken from the composites N-I for only sintering conditions. (**b**) Low cycle fatigue compression test taken from the composites N-I for only sintering conditions. (**c**) Static compression test and incremental test results carried out for the specimens taken from the composites N-II for only sintering conditions. (**d**) Low cycle fatigue compression test taken from the composites N-II for only sintering conditions. (**e**) Static compression test for the sintering (left) and sintering + forging (right) conditions taken from the composites N-III. (**f**) Low cycle fatigue compression test taken from the composites N-III for only sintering conditions

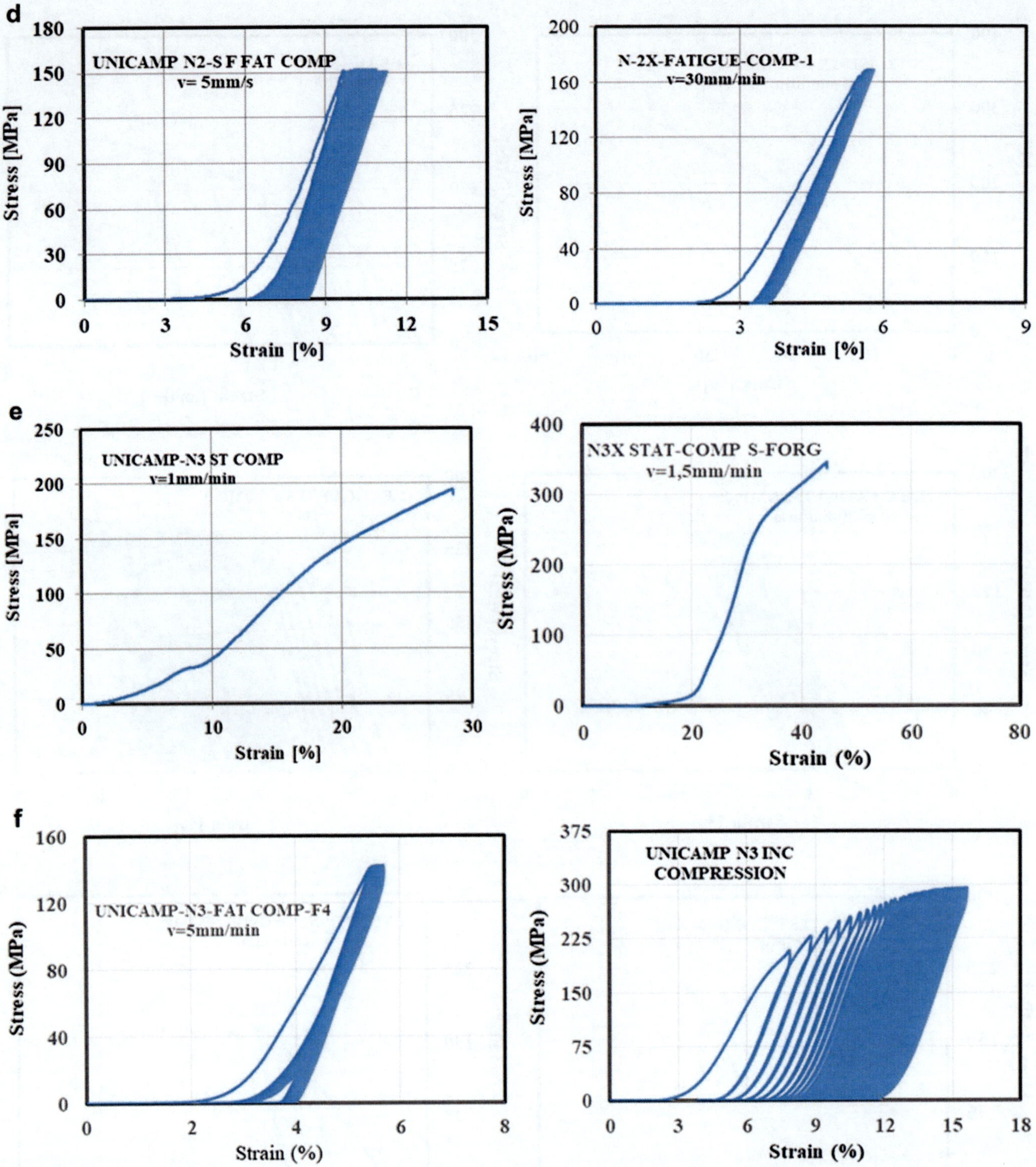

Fig. 5 (continued)

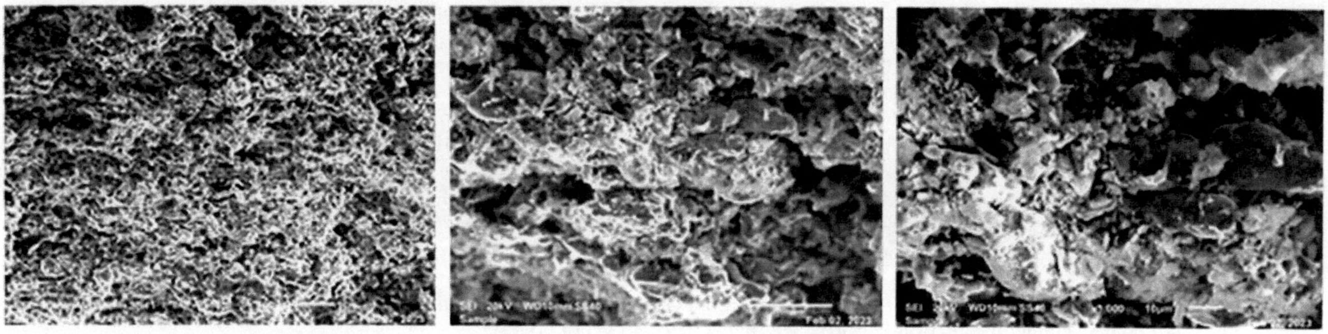

Fig. 6 Fracture surfaces taken from different specimens of the second composition, N-II, produced for only sintering conditions

Table 3 Comparison of the static compression test results with the two-part process

Composite name	UCS (MPa) Sintering (mean values)	UCS (MPa) Sinter + forging (not broken)
N-I	255 ± 15	480
N-II	250 ± 10	490
N-III	300 ± 10	490

Table 4 Fatigue compression test results of the three composites given for the sintered and sintered + -forged specimens

Maximum fatigue stress (mean values) without failure at 10,000 cycles	N-I σ_{max} (MPa)	N-II σ_{max} (MPa)	N-III σ_{max} (MPa)
Sintering	220	180 ± 15	200 ± 10
Sinter + forging	370	300	310

Conclusions

Innovative hybrid aluminium matrix composites (AA7075 + A1050) were designed from the recycled chips supplied by a Brazilian aeronautical company, through the combined method of powder metallurgy followed by "sintering + forging". In the present work, three different composites were designed. As the primary reinforcement, very fine rice husk was used and as the secondary reinforcements those based on Cu/Mo/GNPs were used as a low-cost manufacturing process for aeronautical applications.

In general, the static and incremental cyclic compression tests and also the LCF tests give encouraging results. For this common research project, operational parameters for the static and cyclic tests should be taken into account to optimize the composition with the reinforcement levels.

Acknowledgements The authors acknowledge financial support from CNPq – National Council for Scientific and Technological Development, (Brazil) and the Program French Cathedra UNICAMP/French Consulate in São Paulo, Brazil. The third author, E. Bayraktar, personally thanks the French Consulate in São Paulo for partial financial support in this research project. We acknowledge the materials laboratory, University of São Carlos, Brazil, for technical help and for the gas atomization process of our fresh scrap aluminium chips.

References

1. Alaneme, K.K., Akintunde, I.B., Olubambi, P.A., Adewale, T.M.: Fabrication characteristics and mechanical behaviour of rice husk ash–alumina reinforced Al–Mg–Si alloy matrix hybrid composites. J. Mater. Res. Technol. **2**, 60–67 (2013)
2. Gatamorta, F., Enginsoy, H.M., Bayraktar, E., Miskioglu, I., Katundi, D.: Design of recycled Alumix-123 based composites reinforced with γ-Al_2O_3 through combined method; sinter + forging. In: Mechanics of Composite, Hybrid and Multifunctional Materials, vol. 6, pp. 9–17. Springer, Cham (2021)
3. Alaneme, K.K., Adewale, T.M., Olubambi, P.A.: Corrosion and wear behaviour of Al–Mg–Si alloy matrix hybrid composites reinforced with rice husk ash and silicon carbide. J. Mater. Res. Technol. **3**(1), 9–16 (2014)

4. Ravi Kumar, K., Pridhar, T., Sree Balaji, V.S.: Mechanical properties and characterization of zirconium oxide (ZrO_2) and coconut shell ash (CSA) reinforced aluminium (Al 6082) matrix hybrid composite. J. Alloys Compd. **765**, 171–179 (2018)
5. Madakson, P.B., Yawas, D.S., Apasi, A.: Characterization of coconut shell ash for potential utilization in metal matrix composites for automotive applications. Int. J. Eng. Sci. Technol. **4**, 1190–1198 (2012)
6. Enginsoy, H., Gatamorta, F., Bayraktar, E., Robert, M., Miskioglu, I.: Experimental and numerical study of Al-Nb_2Al composites via associated procedure of powder metallurgy and thixoforming. Composites Part B. **162**, 397–410 (2019). https://doi.org/10.1016/j.compositesb.2018.12.138
7. Kurşun, A., Bayraktar, E., Enginsoy, H.M.: Experimental and numerical study of alumina reinforced aluminium matrix composites: processing, microstructural aspects and properties. Composites Part B. **90**, 302–314 (2016)
8. Xu, Q., Qu, S.: Irreversible deformation of metal matrix composites: a study via the mechanism-based cohesive zone model. Mech. Mater. **89**, 72–84 (2015)
9. Breslin, M.C., Ringnalda, J., Xu, L., Fuller, M., Seeger, J., Daehn, G.S., Otani, T., Fraser, H.L.: Processing, microstructure, and properties of co-continuous alumina-aluminum composites. Mater. Sci. Eng. A. **195**, 113–119 (1995)
10. Mihlyuzova, L., Enginsoy, H.M., Slavov, S., Dontchev, D., Bayraktar, E.: Experimental and finite element study of recycled aluminium (AA7075) matrix composites reinforced of TiC/MoS_2/γ-Al_2O_3Fibre/Nb_2Al. In: Mechanics of Composite, Hybrid and Multifunctional Materials, Fracture, Fatigue, Failure and Damage Evolution, vol. 3, pp. 41–47. Springer, Cham (2022)
11. Ezeddini, S., Katundi, D., Bayraktar, E., Miskioglu, I.: Laser cutting of the TiN+Al_2O_3 reinforced aluminium matrix composites through semisolid sintering. In: Mechanics of Composite and Multi-functional Materials, vol. 6, pp. 115–129. Springer, Cham (2017)
12. Zebarjad, S.M., Sajjadi, S.A.: Microstructure evaluation of Al-Al2O3 composite produced by mechanical alloying method. Mater. Des. **27**, 684–688 (2006)
13. Zebarjad, S.M., Sajjadi, S.A.: Dependency of physical and mechanical properties of mechanical alloyed Al-Al2O3 composite on milling time. Mater. Des. **28**, 2113–2120 (2007)

Development of Ni-Al-Based Composites Reinforced with Recycled AA7075 + AA1050 and Ceramics Produced by the Sintering + Forging Process

Fabio Gatamorta, Olga Klinkova, Özgür Aslan, Ibrahim Miskioglu, and Emin Bayraktar

Abstract In this study, the microstructural formation and static/cyclic compression behavior of "Ni-Al+AA7075 +AA1050"-based composites reinforced with ceramics (TiC-TiB_2) have been evaluated. It is aimed at creating a new design to be an alternative to traditional alloys/composites used in the aeronautical industry. These composites are generally produced using a combined method that we call "sinter + forging processes". The static and dynamic properties and also the microstructure (including the distribution of reinforcement elements) are evaluated in detail.

Keywords Recycled composites · Static-cyclic time depending tests · Ceramics

Introduction

The improvement of nickel and aluminium intermetallics-based composites reinforced with recycled aluminium ndA1050 + AA7075 now constitutes a new idea for creating an economical composite design. These composites are now extensively found in manufacturing engineering owing to their strength properties. Applications for these composites are found to adapt very well to the manufacturing of the parts used in thermo-compressors for aircraft engineering [1–8]. Some of the requests for these pieces have remarkable mechanical properties (static, cyclic, time dependent, etc.) and also a healthy microstructure. In fact, these new composites [1–7] also have multifunctional properties. The mechanical and metallurgical properties of these materials have been studied in details by using static and cyclic/time-dependent mechanical tests. In this work, new innovative composites have been developed by using different fine ceramic reinforcements such as TiB_2 and TiC [6–14]. Our research project has shown that the reinforcements of the fine TiB_2, TiC (1–10 μm) ceramic particles, minor nano fine copper and nano graphene nanoplatelets (GNPs) can improve the mechanical properties of these composites very much [5–9]. For this reason, a special combined method called "sintering + forging" was used for the production of these composites.

In the present work, the static and dynamic behaviours of these composites were evaluated to compare them with those described in previous studies carried out by the same research group.

F. Gatamorta
State University of Rio de Janeiro, Rio de Janeiro, Brazil
e-mail: fabio.gatamorta@uerj.br

O. Klinkova
Isae-Supmeca-Paris, Paris, France
e-mail: olga.klinkova@isae-supmeca.fr

Ö. Aslan
Atilim University, Department of Computational Mechanics, Ankara, Turkey
e-mail: ozgur.aslan@atilim.edu.tr

I. Miskioglu
Michigan Technological University, ME-EM Department, Houghton, MI, USA

E. Bayraktar (✉)
Isae-Supmeca-Paris, Paris, France
e-mail: emin.bayraktar@isae-supmeca.fr

F. Gardea et al. (eds.), *Mechanics of Composite, Hybrid and Multifunctional Materials, Volume 5*,
Conference Proceedings of the Society for Experimental Mechanics Series, https://doi.org/10.1007/978-3-031-50478-5_6

Experimental Conditions

As a matrix for these composites, "Ni-Al + A1050 + AA7075" was chosen from the recycled chips supplied by a Brazilian aeronautic company. Recycled aluminium (AA7075) chips were gas atomized after mixing with pure A1050 (50 wt%), which was prepared by high-energy milling in a planetary ball mill in an inert argon atmosphere to prevent oxidation of the powders (20/1 ball/powder ratio). Ni-Al powder was obtained from the WDR chemical society with a particle size of 1–5 μm. Particle size of the final mixture of A1050 and AA7075 was measured with laser granulometry (from 15 up to 20 μm) for the homogenous mixture. Zinc stearate was used as a lubricant during the preparation of the composite. The thermal behaviour of AA7075 powder was evaluated using differential scanning calorimetry thermogravimetric analysis (DSC-TGA). All of the details of the experimental conditions were given in previously published papers [10, 12].

This matrix was divided in four parts and reinforced with TiB_2 and TiC and as minor reinforcements, fine copper and nano graphene were used respectively. During the milling, pure nano AA1050 (3–5 wt%) was also added to facilitate and homogenize the mixture of two types of recycled aluminium. Sintering was performed under argon gas in the oven. At the first stage of sintering, compacted samples were heated at 350 °C for 45 min to eliminate lubricant and other artefacts. At the second stage of sintering, heating was conducted up to the maximum sintering temperature (630 °C). After that, hot forging was carried out to complete the manufacturing processes at a lower temperature than the sintering temperature. Mechanical and physical properties were evaluated through microhardness, and static-cyclic-time-dependent fatigue compression tests were carried out in order to understand the toughening mechanism of these composites under laboratory conditions at the university of Campinas, UNICAMP/FEM materials laboratory.

Results and Discussion

Figure 1 shows the DSC diagram measured for A7075 alloy and also simulation of the fraction of solid depending on the temperature calculated using the software "Thermo-Calc" in the matrix to determine the critical transformation points during the heating and cooling stages. Additional information was given for energy-dispersive spectroscopy (EDS) chemical analysis in the next sessions for each composite.

In the following tables, the chemical composition of the A7075 aluminium alloy (Table 1) and the general composition of four hybrid composites designed within the frame of our common research project (Table 2).

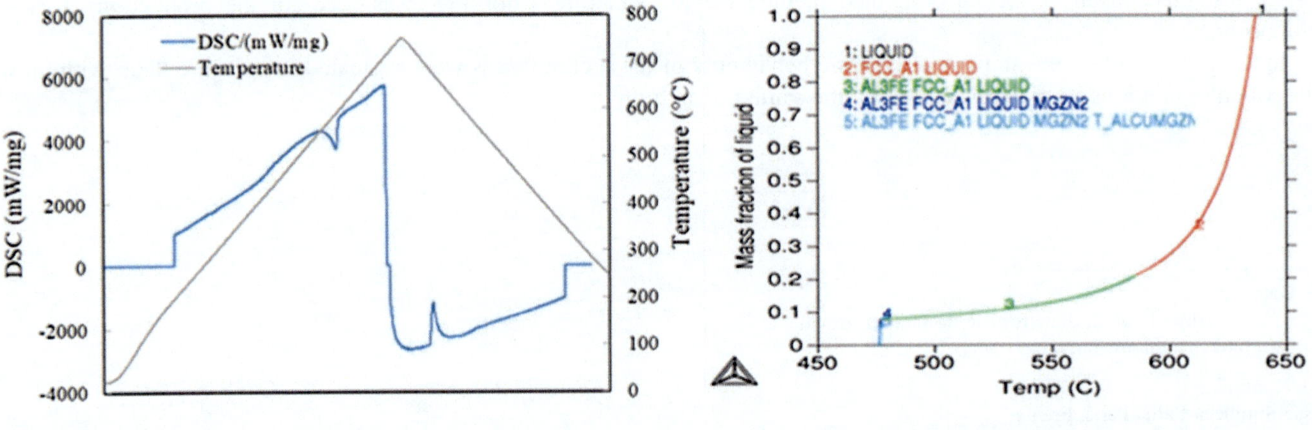

Fig. 1 Differential scanning calorimetry diagram measured for A7075 with a heating rate of 5 °C/min and simulation of the fraction of the solid depending on the temperature [2, 8, 11]

Table 1 Chemical composition of scrap A7075 alloy (wt%)

Element	Al	Cu	Fe	Mg	Mn	Si	Ni	Zn	Cr	Zr
wt%	Balance	1.48	0.23	2.11	0.07	0.10	0.01	5.29	0.22	0.02

Table 2 Compositions of the four composites (wt%)

Composite name	A1050 (50%) AA7075 (50%)	Ni-Al	TiC	TiB_2	Cu	ZnO	GNPs	Zn stearate
E-3	B	15	15	–	2	2	0.25	2
E-6	B	30	15	–	2	2	0.25	2
E-7	B	15	–	15	2		0.25	2
E-8	B	30	–	15	2	2	0.25	2

Table 3 Electrical and thermal properties measured for four composites with microhardness values

Composite name	Electrical conductivity at ambient (S/m)	Thermal conductivity (W/mK)	Microhardness ($HV_{0.1}$)
E-3	5.45×10^9	4.325	125 ± 15
E-6	6.10×10^9	4.870	185 ± 10
E-7	5.15×10^9	4.023	135 ± 15
E-8	5.60×10^9	4.655	140 ± 15

Fig. 2 Sintering process carried out at 630 °C followed by hot forging of the specimens (left) and compression test device adapted on the Zwick test machine (ISAE-SUPMECA/Paris)

For each composite, the Ni-Al intermetallic was reinforced with one ceramic at a constant ratio; only the Ni-Al intermetallic was changed for each group. The addition of minor reinforcements such as copper and GNPs was aimed at increasing the chemical bonding diffusion and strengthening of the composites respectively.

Electrical conductivity levels were measured using an "Agilent 4338B Milli-ohm Meter" in the materials laboratory of ISAE-SUPMECA-Paris. Three specimens were measured for each composite and the mean values are given in Table 3. For the measurements, DC-regulated power supply voltage and current were set as 20 V and 20 A respectively. The data acquisition card "NI9234" was connected in parallel with the output of the power to acquire the voltage data (voltage input accuracy was 24 bits). A high-precision multi-meter "Agilent U1253N" was connected in the series to measure the current intensity (A). For the same specimens, thermal conductivity measurements carried out in our laboratory and microhardness measurements taken from the samples produced under the same conditions as for these composites are also presented in Table 3. All the data for the electrical and thermal measurements were revealed using the LabVIEW program. These results should obviously be assumed to be indicative data under laboratory conditions. It is difficult to measure a constant value, as the structure is heterogenous.

In Fig. 2, a sintering process carried out at 630 °C followed by hot forging of the specimens is presented as well as a compression test device adapted on the Zwick test machine (ISAE-SUPMECA/Paris).

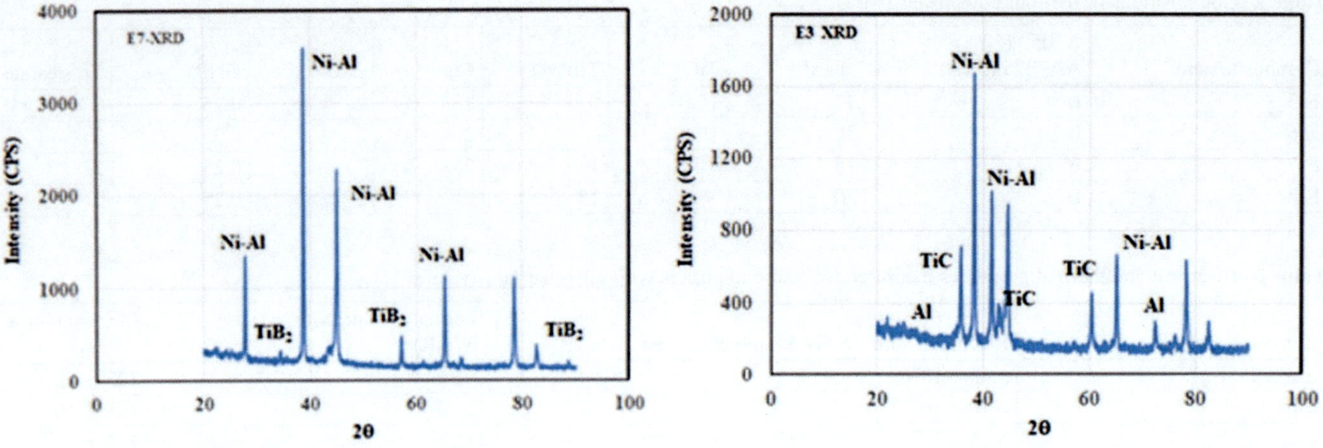

Fig. 3 X-ray diffraction analyses for two Ni-Al-A10506-A7075 composites containing TiB_2 (E7) and TiC (E3)

X-ray diffraction analyses were also carried out at CARAM's Physical Metallurgy Laboratory for the composites to detect the structure for each composite (safety control of each composite). In the present work there were only two different ceramics, called E3 and E7 (Fig. 3).

As for the microstructure of the four composites, a detailed metallographic analysis was carried out by mapping and EDS chemical analyses for each composite designed in the present work. They have shown very close similarity to each other because the structure contains different amounts of Ni-Al. Figure 4 presents this evaluation for only one composite, called E-7. EDS chemical analyses, fracture surface evaluation of the fatigue test and mapping analyses have shown that an homogeneous distribution of the particles in the microstructure.

In general observation of the surface microstructures and mapping of these composites, it seems a homogenous distribution of the reinforced particles in the matrix, even though some agglomerations were observed under laboratory conditions. It means that these composites will be produced soon by our industrial partners at the level of industrial scale.

It seems that the strengthening values should be higher if the minor reinforcements could be dispersed homogenously. Within the framework of this research, these values were obtained under laboratory conditions, which provides a useful tool for industrial applications.

To understand the toughening mechanism in detail, three different comprehensive tests were conducted in the present work. Static and incremental cyclic compression tests were performed to evaluate the mechanical behaviours of these innovative hybrid composites, with a deformation rate of 1.5 mm/min on the Zwick test machine (ISAE-SUPMECA/Paris) with a load cell capacity of 100 kN.

Low-cycle fatigue (LCF) tests were also performed on a closed loop test device with a frequency of 0.5 Hz.

The LCF tests were performed with a computer-controlled adapter with a 100-kN load cell (ASTM E606). These tests were performed under stress-controlled, fully reversed compression-compression loading. LCF tests were limited to 10^4 cycles in this work and the maximum applied fatigue strength was chosen at around the yield strength, between 70% and 80% of the maximum strength.

Figure 5a–d show the total results of the mechanical behaviours of the four composites under the static and incremental cyclic as well as time-dependent low-cycle fatigue conditions respectively.

The static compression test results were compared here for the sintering + forging process. All of the results are summarized in Table 4.

All of the fatigue specimens produced from four composites have shown ductile transgranular fatigue damage behaviour, most of them in the matrix, and with relatively fine grains. The cohesion of the reinforcements with the matrix are very strong and distribution was homogenous for each composite. In general, the integration of ceramic particle reinforcements into the metal matrix can sometimes decrease the ductility and toughness levels of the metal matrix composites, but very often all four composites have shown ductile fracture behaviour. As we have indicated in the previous section, these results are only limited to those performed under the laboratory conditions. Table 5 presents the total fatigue test results. All of the results were given as the mean values.

It is summarized here that the test results carried out on the four composites show ductile behaviour that is related to the tough and sound microstructure, with strong cohesion of the reinforcements with the matrix.

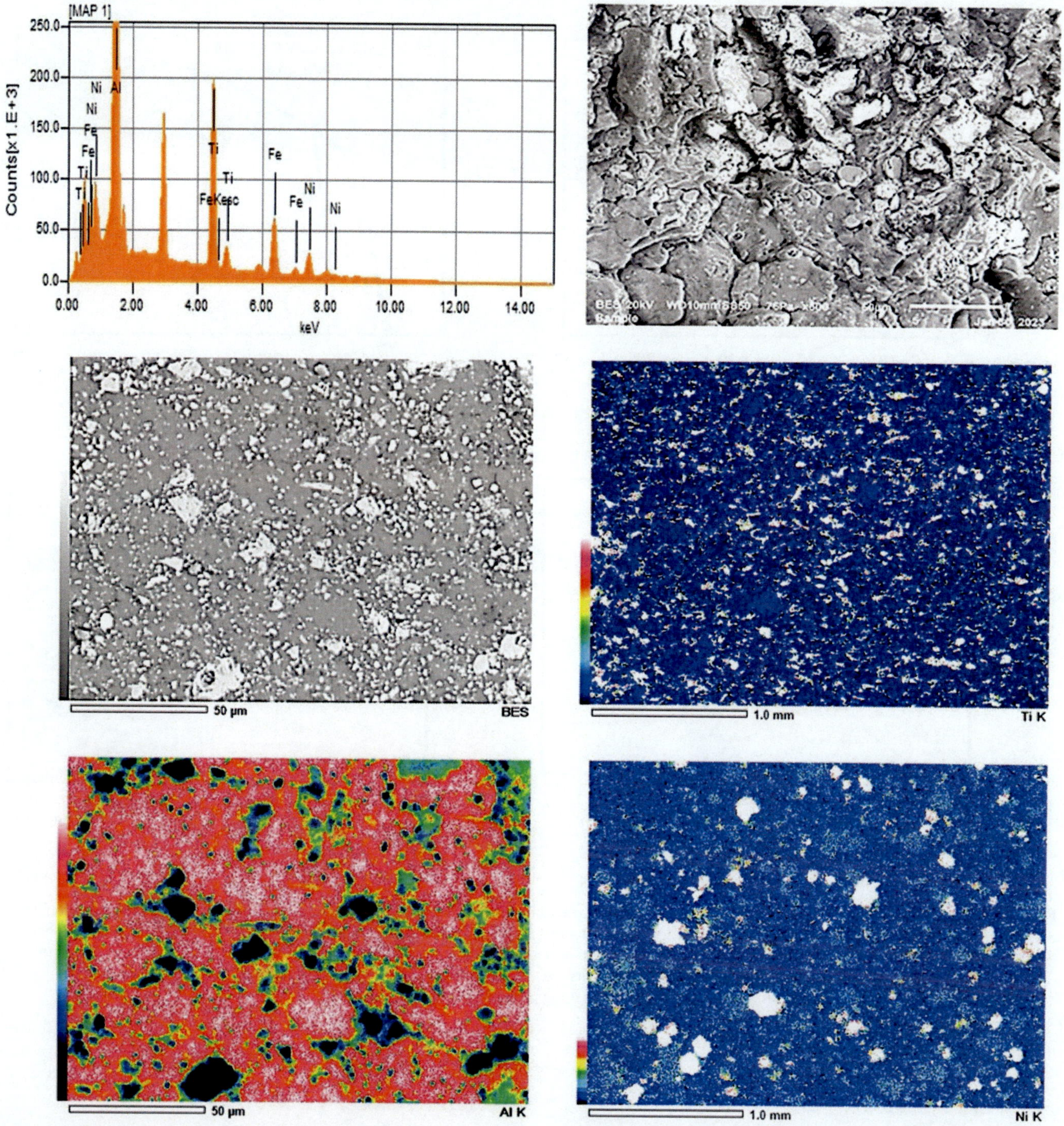

Fig. 4 Energy-dispersive spectroscopy chemical analyses, fracture surfaces at the fatigue test, mapping analyses of the composite E7

Conclusions

Innovative multifunctional "NiAl-A1050-AA7075" matrix composites (with an equal mixture of AA7075 + A1050) were designed from the recycled chips supplied by a Brazilian aeronautical company, through the combined method of powder metallurgy followed by "sintering + forging". In the present work, four different composites were designed. As the primary reinforcements, fine ceramics was used and as secondary reinforcements those based on Cu and GNPs were used as a low-cost manufacturing process for use in aeronautical applications.

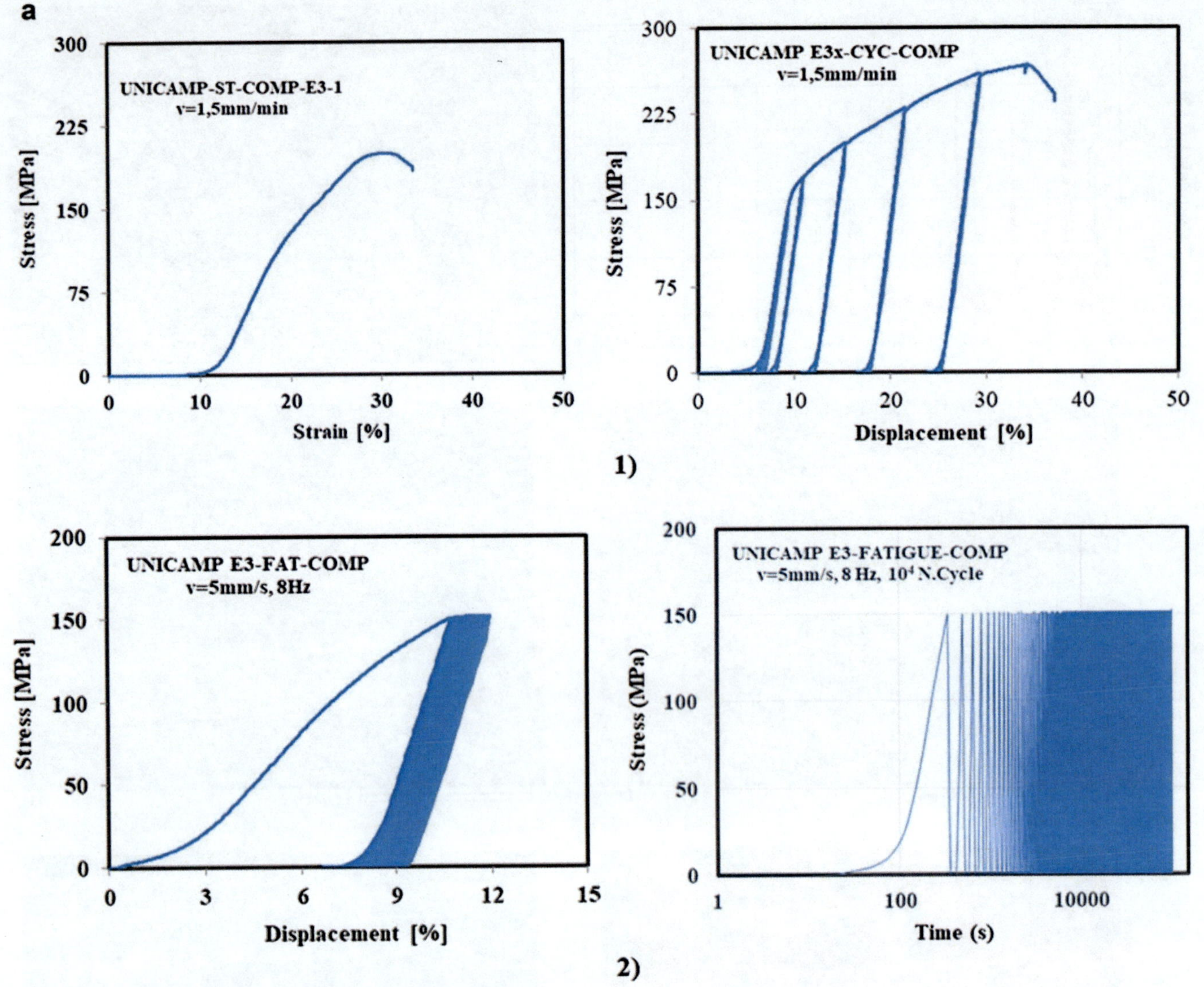

Fig. 5 (**a**) Static compression test and incremental test results carried out for the specimens taken from the composite E-3 and low-cycle fatigue compression test results. (**b**) Static compression test and incremental test results carried out for the specimens taken from the composite E-6 and the low-cycle fatigue compression test results. (**c**) Static compression test and incremental test results carried out for the specimens taken from the composite E-7 and the low-cycle fatigue compression test results. (**d**) Static compression test and incremental test results carried out for the specimens taken from the composite E-8 and the low-cycle fatigue compression test results

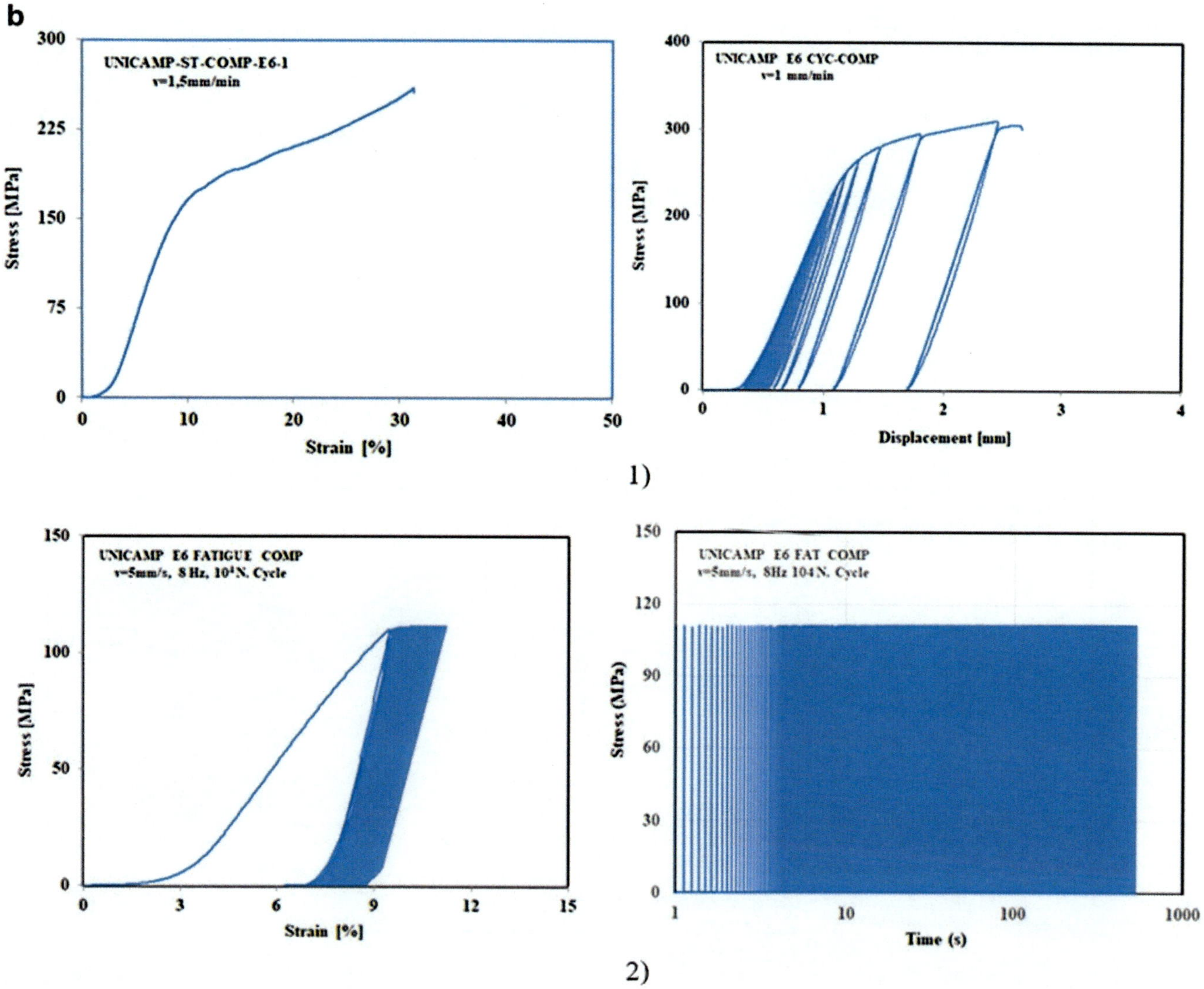

Fig. 5 (continued)

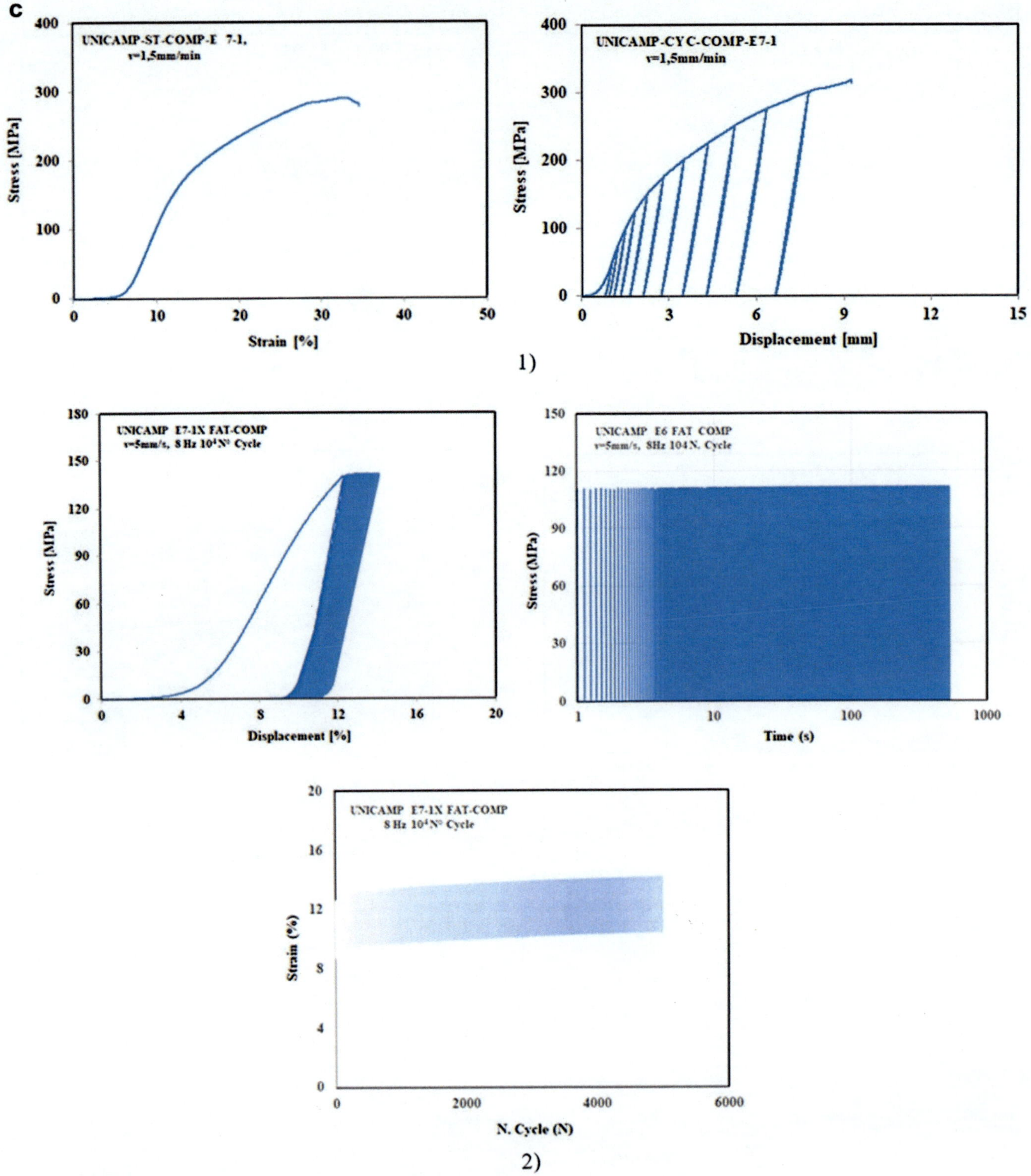

Fig. 5 (continued)

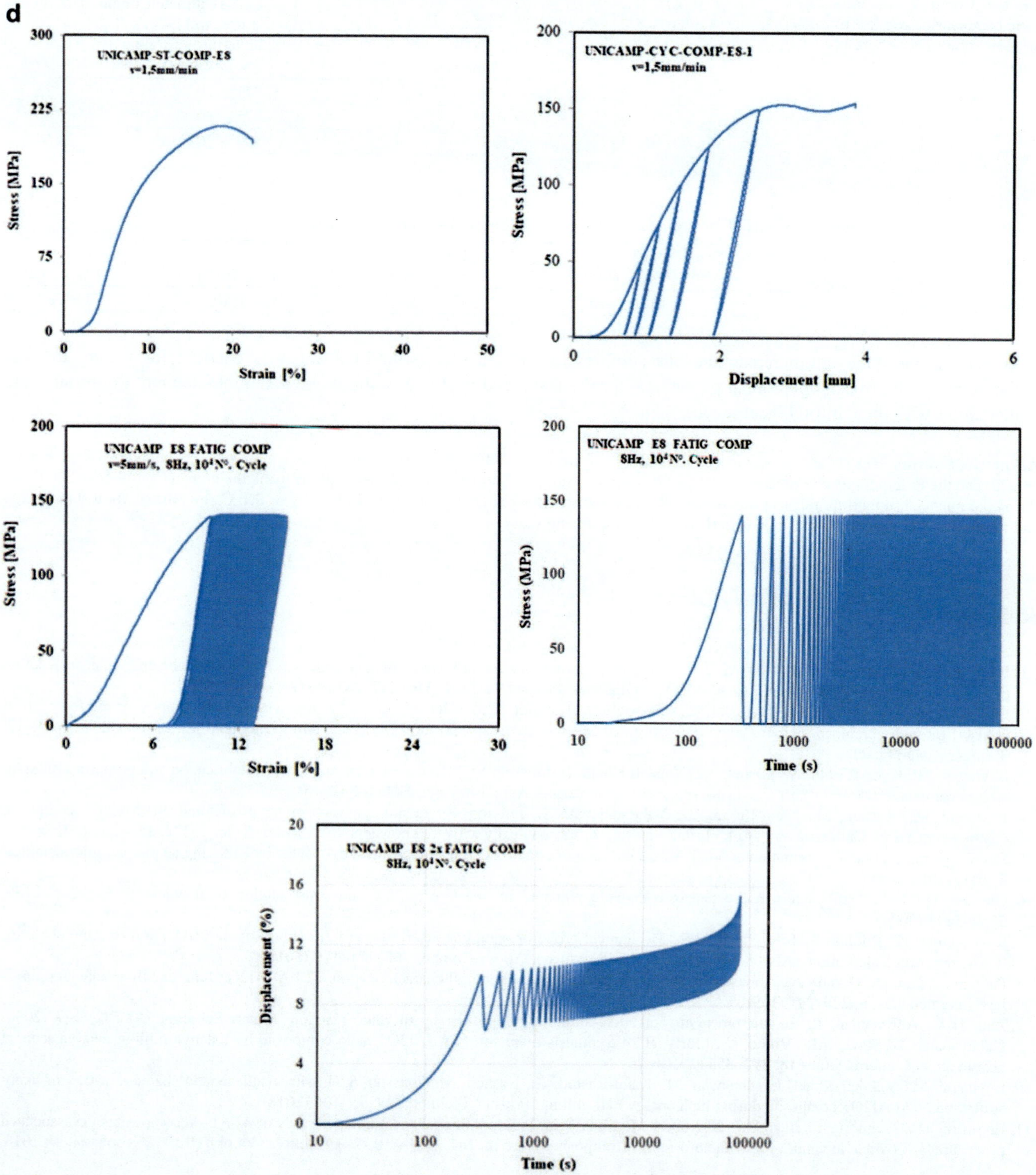

Fig. 5 (continued)

Table 4 Comparison of the static compression test results for each composite as a mean value

Composite name	UCS (Ultimate Compression Stress at fracture) (MPa) Sintering + forging (mean values)
E-3	275 ± 10
E-6	275 ± 10
E-7	300 ± 10
E-8	250 ± 10

Table 5 Fatigue compression test results of the four composites given for sintered + forged specimens

Maximum fatigue stress (mean values) without failure at 10,000 cycles	E-3 σ_{max} (MPa)	E-6 σ_{max} (MPa)	E-7 σ_{max} (MPa)	E-8 σ_{max} (MPa)
Sintering + Forging	150 ± 10	130 ± 15	150 ± 10	150 ± 15

In general, the static and incremental cyclic compression tests as well as the LCF tests give encouraging results. For this common research project, operational parameters for the static and cyclic tests should be taken into account to optimize the composition with the reinforcement levels.

Acknowledgements The authors acknowledge financial support from the CNPq – National Council for Scientific and Technological Development, (Brazil) and the Program French Cathedra UNICAMP/French Consulate in São Paulo, Brazil. The authors thank the French Consulate in São Paulo for partial financial support for this research project. We acknowledge the materials laboratory, University of São Carlos, Brazil, for technical help and for the gas atomization process of our fresh scrap aluminium chips.

References

1. Beyhaghi, M., Kiani-Rashid, A., Khaki, J.V., Kashefi, M., Jonsson, S.: Influences of mechanical activation and heating rate on reaction processes in combustion synthesis of NiAl-Al2O3 composites. Powder Technol. **346**, 237–247 (2019)
2. Gatamorta, F., Enginsoy, H.M., Bayraktar, E., Miskioglu, I., Katundi, D.: Design of recycled Alumix-123 based composites reinforced with γ-Al_2O_3 through combined method; sinter + forging. In: Mechanics of Composite, Hybrid and Multifunctional Materials, vol. 6, pp. 9–17. Springer, Cham (2021)
3. Beyhaghi, M., Kiani-Rashid, A., Kashefi, M., Vahdati Khaki, J., Jonsson, S.: Effect of powder reactivity on fabrication and properties of NiAl/Al2O3 composite coated on cast iron using spark plasma sintering. Appl. Surf. Sci. **344**, 1–8 (2015)
4. Beyhaghi, M., Kashefi, M., Kiani-Rashid, A., Vahdati Khaki, J., Jonsson, S.: In-situ synthesis of nanostructured NiAl-Al2O3 composite coatings on cast iron substrates by spark plasma sintering of mechanically activated powders. Surf. Coat. Technol. **272**, 254–267 (2015)
5. Lin, C., Hong, S., Lee, P.: Formation of NiAl-Al2O3 intermetallic-matrix composite powders by mechanical alloying technique. Intermetallics. **8**, 1043–1048 (2000)
6. Oleszak, D.: NiAl-Al2O3 intermetallic matrix composite prepared by reactive milling and consolidation of powders. J. Mater. Sci. **39**, 5169–5174 (2004)
7. Ravi Kumar, K., Pridhar, T., Sree Balaji, V.S.: Mechanical properties and characterization of zirconium oxide (ZrO_2) and coconut shell ash (CSA) reinforced aluminium (Al 6082) matrix hybrid composite. J. Alloys Compd. **765**, 171–179 (2018)
8. Padmavardhani, D., Gomez, A., Abbaschian, R.: Synthesis and microstructural characterization of NiAl-Al2O3 functionally gradient composites. Intermetallics. **6**, 229–241 (1998)
9. Zhu, H.X., Abbaschian, R.: In-situ processing of NiAl–alumina composites by thermite reaction. Mater. Sci. Eng. A. **282**, 1–7 (2000) Udhayabanu, V., Ravi, K.R., Vinod, V., Murty, B.S.: Synthesis of in-situ NiAI-Al2O3 nano composite by reactive milling and subsequent heat treatment. Intermetallics **18**, 353–358 (2010)
10. Beyhaghi, M., Vahdati Khaki, J., Manawan, M., Kiani-Rashid, A., Kashefi, M., Jonsson, S.: In-situ synthesis and characterization of nano-structured NiAl-Al2O3 composite during high energy ball milling. Powder Technol. **329**, 95–106 (2018)
11. Enginsoy, H., Gatamorta, F., Bayraktar, E., Robert, M., Miskioglu, I.: Experimental and numerical study of Al-Nb_2Al composites via associated procedure of powder metallurgy and thixoforming. Composites Part B. **162**, 397–410 (2019). https://doi.org/10.1016/j.compositesb.2018.12.138
12. Kurşun, A., Bayraktar, E., Enginsoy, H.M.: Experimental and numerical study of alumina reinforced aluminium matrix composites: processing, microstructural aspects and properties. Composites Part B. **90**, 302–314 (2016)
13. Breslin, M.C., Ringnalda, J., Xu, L., Fuller, M., Seeger, J., Daehn, G.S., Otani, T., Fraser, H.L.: Processing, microstructure, and properties of co-continuous alumina-aluminum composites. Mater. Sci. Eng. A. **195**, 113–119 (1995)
14. Mihlyuzova, L., Enginsoy, H.M., Slavov, S., Dontchev, D., Bayraktar, E.: Experimental and finite element study of recycled aluminium (AA7075) matrix composites reinforced of TiC/MoS_2/γ-Al_2O_3Fibre/Nb_2Al. In: Mechanics of Composite, Hybrid and Multifunctional Materials, Fracture, Fatigue, Failure and Damage Evolution, vol. 3, pp. 41–47. Springer, Cham (2022)

Design of Recycled "A356-A7075" Matrix Composites Reinforced with "Nb_2Al-Zr" Produced by Sinter Forging

Fabio Gatamorta, Dhurata Katundi, Olga Klinkova, Emin Bayraktar, and Ibrahim Miskioglu

Abstract In this research, the mechanical properties of recycled aluminum (A356-A7075)-based composites reinforced with Nb_2Al-ZrO_2-TiAl have been evaluated. Different amounts of Nb_2Al doped with ZrO_2 were mixed with atomized recycled aluminum matrix containing "AA356-A7075" and processed by ball milling. The mixture was then sintered followed by hot forging. Static and cyclic stress relaxation compression tests and low velocity impact tests were carried out to evaluate damage behavior. Interface and microstructure of these composites were also evaluated by scanning electron microscope (SEM).

Keywords Recycled hybrid composite · Nb_2Al-Zr · Static-cyclic tests

Introduction

Aluminum matrix composites, especially recycled fresh scraps, are recently used in the aerospace and automobile industries for structural parts. Fresh scraps are mainly gas atomized [1–8] in order to prepare fine matrix for composites. The aim of using nanoparticles in very low percentage and microscale reinforcements in higher percentage is to increase the durability of the material under extensive service conditions such as static and cyclic loading, very high friction, and vibration. Nanoparticle reinforcements have more advantages in mechanical strength and higher resistance against wear than micron-sized reinforcements. As a result, they are very often used in the manufacturing of different parts in new composites design [7–16]. For this reason, we have used the triple effect of the large amount of micro reinforcements together with nanoparticles in small quantities. Solid-state process was used in the manufacturing of all the samples.

Our research groups UNICAMP-FEM, ISAE-SUPMECA, Michigan Tech, USA, have developed an innovative process to improve the mechanical properties and increase the toughening mechanisms of recycled aluminum composites reinforced with micro- and nanoparticles (Nb_2Al/ZrO_2/GNPs) [1, 4, 11, 12]. This study concentrates on the toughening mechanism conducted on the mechanical properties, microstructural characterization, in general meaning, strengthening mechanisms of the new hybrid aluminum-based composites.

F. Gatamorta
State University of Rio de Janeiro, Rio de Janeiro, Brazil

D. Katundi · O. Klinkova
Isae-Supmeca-Paris, School of Mechanical and Manufacturing Engineering, Saint-Ouen-sur-Seine, France

E. Bayraktar
Unicamp, University of Campinas, FEM, SP, Campinas, Brazil

Isae-Supmeca-Paris, School of Mechanical and Manufacturing Engineering, Saint-Ouen-sur-Seine, France

I. Miskioglu (✉)
Michigan Technological University, ME-EM Department, Houghton, MI, USA
e-mail: imiski@mtu.edu

F. Gardea et al. (eds.), *Mechanics of Composite, Hybrid and Multifunctional Materials, Volume 5*,
Conference Proceedings of the Society for Experimental Mechanics Series, https://doi.org/10.1007/978-3-031-50478-5_7

Experimental Conditions

Two recycled aluminum alloy (AA356 + A7075) powders were used as the matrix for the composites in study. The recycled aluminum, in the form of chips, was supplied by the Brazilian aeronautic company. First, recycled aluminum (AA 351) and AA7075 chips were gas atomized and then they were mixed (50 wt% each) by high energy milling in a planetary ball mill under inert argon atmosphere to prevent oxidation of the powders (20/1 ball/powder ratio). Additionally, 4 wt% of zinc stearate was used as a lubricant during the preparation of the composite. After milling operation, the thermal behavior of the powder of both aluminum alloys (A356–AA7075) was evaluated with DSC-TGA and XRD. The details of these experiments were given in the former papers [1, 4, 11].

The matrix was divided in four parts and reinforced with Nb_2Al, TiAl, and ZrO_2, and as minor reinforcements, molybdenum and copper (Mo 4 wt%, Cu 4 wt%, and GNPs 0.75 wt%) were used. During the milling process, pure nano AA1050 (3–5 wt%) was added to homogenize the mixture of two types of the recycled aluminum alloys. Sintering was performed under argon gas. In the first stage of sintering, compacted samples were heated to 300 °C for 30 min for eliminating lubricant and other artifacts. In the second stage of sintering, heating was conducted up to the maximum sintering temperature (600 °C) for 45 min. In the final stage, hot forging was carried out to complete the manufacturing processes at a lower temperature than the maximum sintering temperature. Mechanical and physical properties were evaluated through micro-hardness and static-cyclic compression tests as well as low velocity impact/drop weight tests were carried out only at one energy level.

Results and Discussion

Figure 1a shows a Differential Scanning Calorimetry (DSC) diagram measured for A350 and A7075 alloys and also simulation of fraction of solid depending on the temperature calculated with software "Thermo-Calc" in the matrix to determine the critical transformation points during the heating and cooling stages. Additional information is given for "EDS" chemical analysis in the next sections for each composite.

After determining the sintering conditions from the DSC analyses, XRD analyses have been carried out on the final composite specimens for the second composition called here JUN-II. The aim of these analyses is to find correct design of the composites and determine the phases on the sintered composites. Figure 2 shows the XRD analyses for only one specimen (JUN-II) to detect the phase information after sintering process of the composite.

Table 1 presents the compositions of the four composites formulated for the innovative hybrid composite designed with two aluminum alloys A356 and A7075 as matrix. Besides the major reinforcements (Nb_2Al, ZrO_2, and TiAl), small amounts of Mo-Cu-GNPs increase strengthening of the composites, thereby generating a strong cohesion of the reinforcements with the matrix mainly on the grain boundaries. For the fine distribution of the reinforcements and to obtain a fine grain size, Cu and nano Mo (<1 μ) and nano graphene platelets (GNPs, 500 m^2/gr with surface particle) were added to the matrix for each composite. Additionally, the presence of Mo and GNPs in the structure increases the mechanical resistance for toughening mainly due to a strong cohesion by chemical diffusion bonding at the interface between the matrix and the reinforcements. We know that fine copper particles added in the composition improve the chemical diffusion bonding in the matrix. The formation of the chemical diffusion bonding mechanism that will be presented in the next section is only indicative and should be improved with new measurements during the course of this research project.

Electrical conductivity levels were measured with an "Agilent 4338B Milli-ohm Meter." Three specimens were measured for each composite, and then, the mean values are given in Table 2. For the measurements, DC-regulated power supply voltage and current were set as 20 V and 20 A, respectively. Data acquisition Card "NI9234" was connected in parallel with the output of the power to acquire the voltage data (voltage input accuracy was 24 bits). A high precision multi-meter "Agilent U1253N" was connected in series to measure the current intensity (*A*). For the same specimens, thermal conductivity measurements carried out in our laboratory and also micro-hardness measurements taken from the samples produced under the same conditions for these composites are also presented in Table 2. All the data for the electrical and thermal measurements were revealed with LabVIEW program. These results obviously should be assumed an indicative data under the laboratory conditions.

In Fig. 3, a sintering process carried out at 600 °C followed by hot forging of the specimens is presented and also compression test device adapted on the Zwick test machine (ISAE-SUPMECA/Paris)

To understand the toughening mechanism, three different comprehensive tests were conducted on the mechanical properties of the new hybrid aluminum-based composites. Results of these tests, static compression at a test speed of

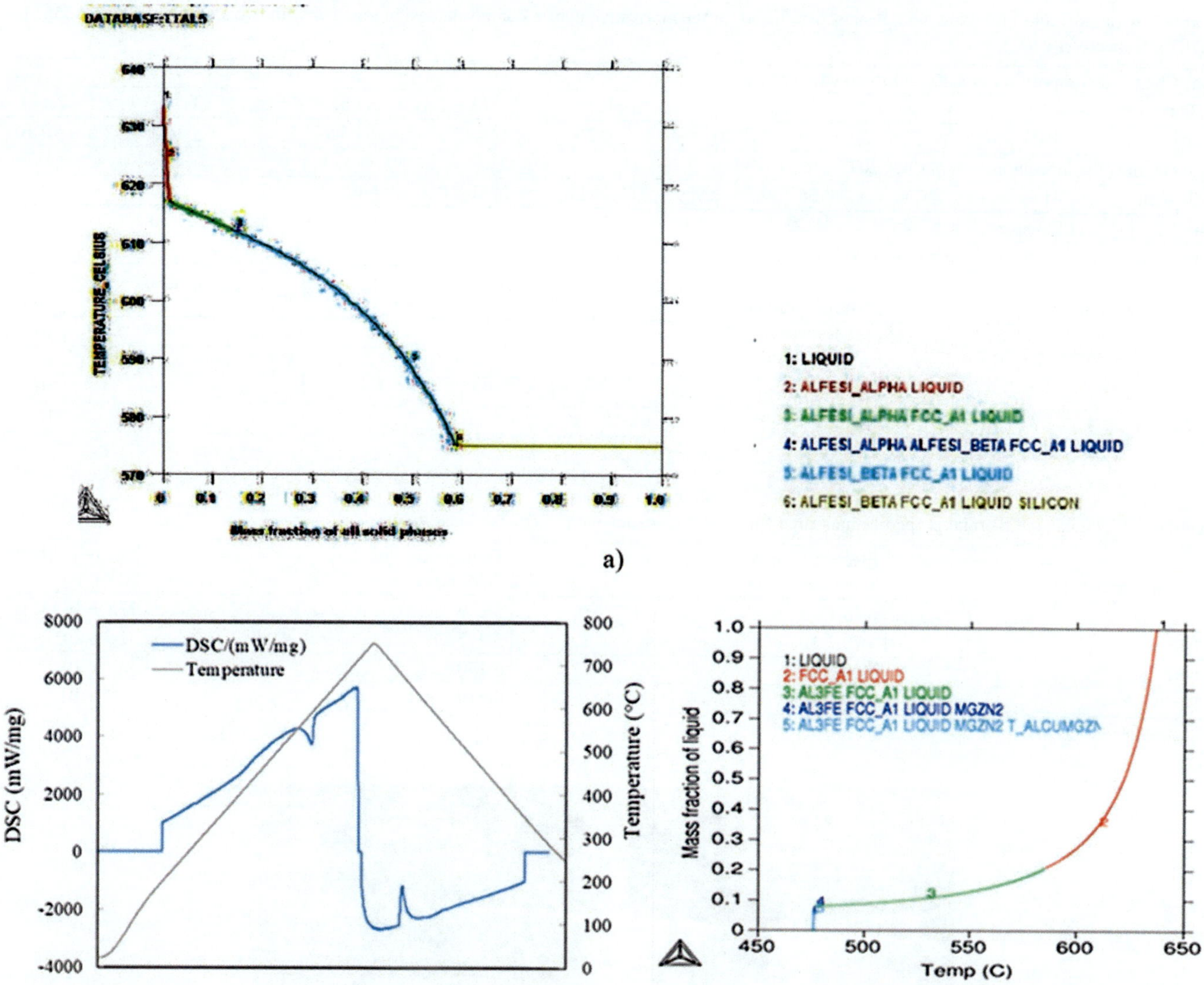

Fig. 1 (**a**) Differential Scanning Calorimetry (DSC) diagram measured for A356 with a heating rate of 5 °C/min and also simulation of the fraction of solid depending on the temperature, (**b**) the same measurements for A7075 and also simulation of the fraction of solid depending on the temperature [*X*]

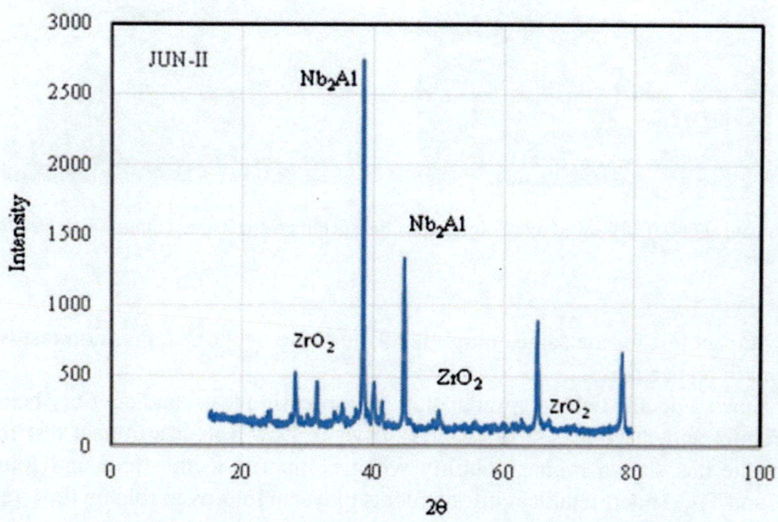

Fig. 2 XRD analyses of the second composite JUN II to detect the phases after sintering of the composite

Table 1 Compositions of the four composites formulated for the innovative hybrid composite design together with two aluminum alloys A356 and A7075, respectively

(a) Chemical composition of scrap A356 powder (wt%)

Element	Al	Cu	Fe	Mg	Mn	Si	Ti	Zn
wt%	Balance	≤0.20	≤0.20	≤0.25	≤0.10	6.5–7.5	≤0.20	≤0.10

(b) Chemical composition of scrap AA 7075 alloy (wt%)

Element	Al	Cu	Fe	Mg	Mn	Si	Ni	Zn	Cr	Zr
wt%	Balance	1.48	0.23	2.11	0.07	0.10	0.01	5.29	0.22	0.02

(c) Compositions of the four composites (wt%)

Composite name	A350 (50%) AA7075 (50%)	Nb_2Al	Ti-Al	ZrO_2
J-I	B	5	5	10
J-II	B	10	5	10
J-III	B	15	5	10
J-IV	B	30	5	10

Table 2 Electrical and thermal properties measured for four composites with micro-hardness values

Composite name	Electrical conductivity at ambient (S/m)	Thermal conductivity (W/mK)	Micro-hardness ($HV_{0,1}$)
J-I	$6.85 * 10^9$	5.650	275 ± 10
J-II	$5.70 * 10^9$	5.255	310 ± 20
J-III	$5.00 * 10^9$	5.130	385 ± 10
J-IV	$3.50 * 10^9$	4.350	410 ± 10

Fig. 3 Sintering process carried out at 600 °C followed by hot forging of the specimens (left) and compression test device adapted on the Zwick test machine (ISAE-SUPMECA/Paris)

1.5 mm/min, low velocity impact test for the same energy level, and also cyclic compression test with a test speed of 5 mm/s are presented in Figs. 4 and 5, respectively.

Four composites have shown a ductile behavior under static compression tests, and ductility behavior decreased slightly as the Nb_2Al intermetallic reinforcement increased to 15 and 30 wt%. Low velocity impact test results have shown similar behavior. The first composite has shown higher ductility when compared to the third and fourth composite series. The combined effect of Nb_2Al and Ti-Al intermetallic reinforcements plays an improved role on the strengthening and toughening

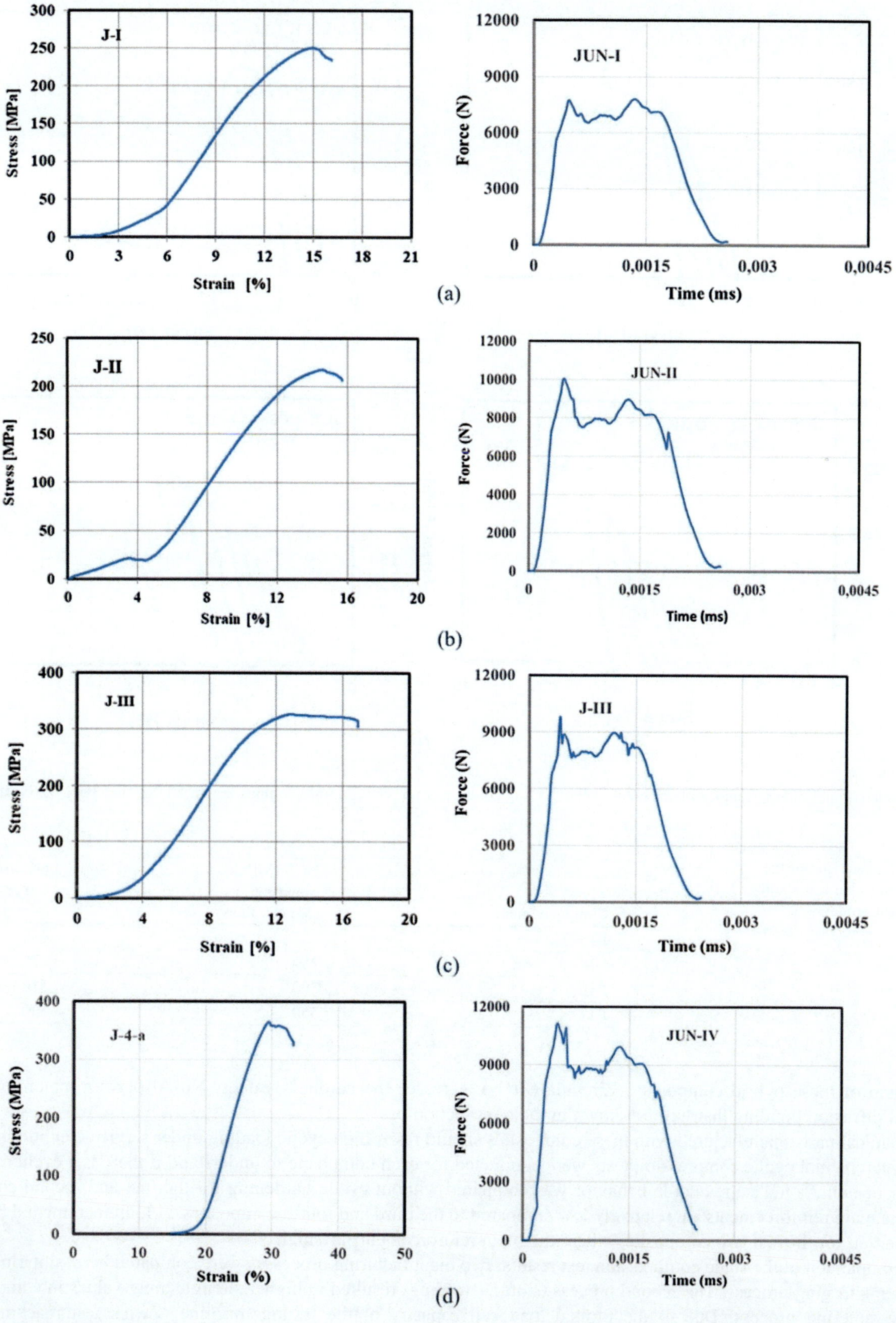

Fig. 4 Static compression test and low velocity impact test carried out for all composites. (**a**) composite J-I, (**b**) composite J-II, (**c**) composites J-III, (**d**) composite J-IV (sintering only)

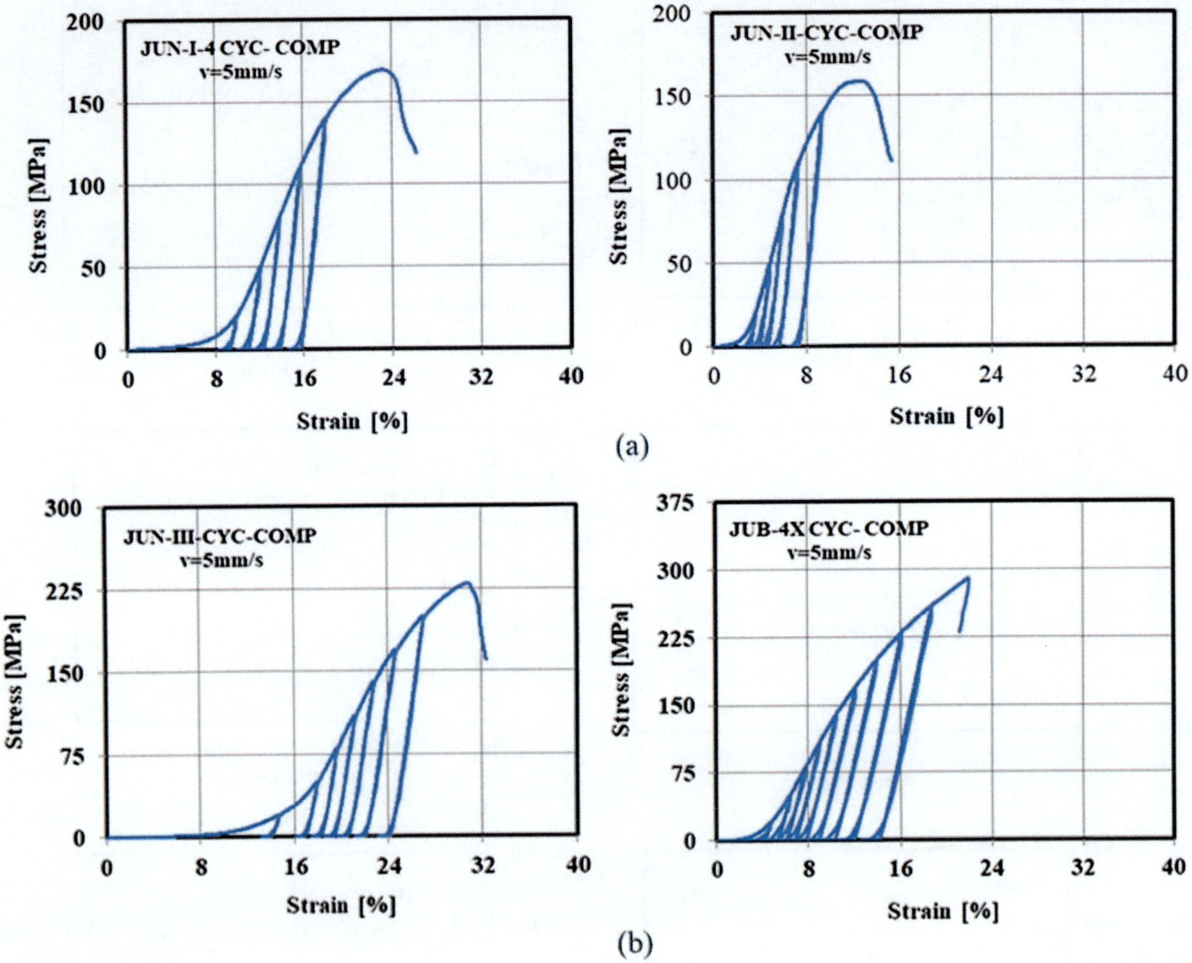

Fig. 5 Static compression test under incremental cyclic loading conditions for sintered only composites. (**a**) JUN-I and JUN II, (**b**) JUN-III and JUN IV

Table 3 Comparison of the static compression test results with the two production processes

Composite name	UCS (MPa) Sintering (Mean values)	UCS (MPa) Sinter + Forging (Not broken)
J-I	235	410
J-II	250	425
J-III	300	510
J-IV	350	530

mechanism of these hybrid composites. Zirconia (ZrO_2) increases the hardness values. Nb_2Al plays a major role on the chemical diffusion bonding that will be shown in the next section.

Mechanical parts manufactured with these composites should resist high cyclic loading under service conditions. For this reason, incremental cyclic compression tests were conducted for each composite to understand if there is a cyclic hardening behavior or not. As the same ductile behavior was compared without cyclic hardening for the first and second composites where the main reinforcements are relatively low compared to the third and fourth composites. A light incremental hardening was observed for the last two composites where the main reinforcements are high.

For comparison of the static compression test results, two manufacturing processes were compared here: sintering process and sinter + forging process. The second process (sinter + forging) resulted in higher strengthening values in comparison to the only sintering process. Due to the limited load cell capacity of the testing machine (Zwick test machine, ISAE-SUPMECA/Paris), specimens produced with sinter + forging process could not be failed. For these specimens, the maximum stress reached during testing is reported in Table 3.

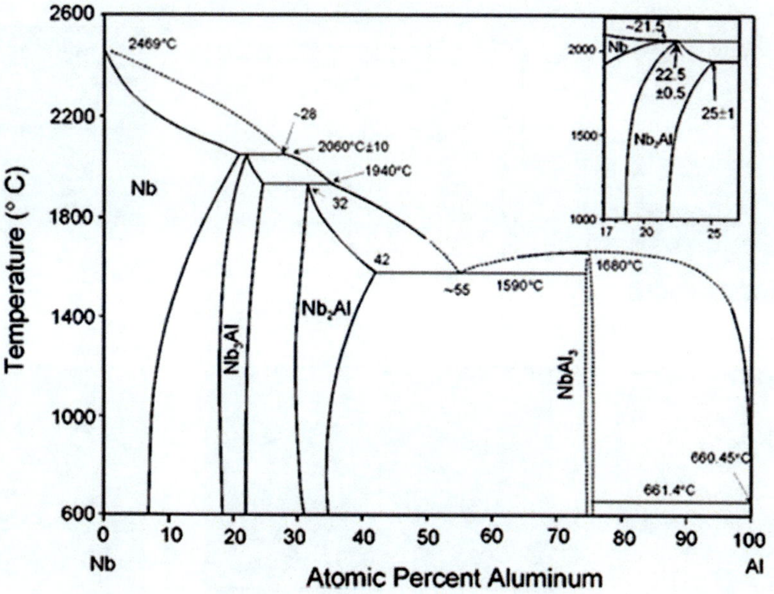

Fig. 6 Binary phase diagram of Nb-Al taken from ASM Metals Handbook [16]

In the final stage, a detailed metallographic analysis were carried out by mapping and EDS chemical analysis for all the composites designed for this study. A binary phase diagram of niobium-aluminum (Nb-Al) is shown in Fig. 6. Based on this diagram, the composites designed in this work are found around certain intermetallic zones (Nb_2Al and Nb_3Al). The main reinforcement, Nb_2Al content was different in the four composites presented here [16].

A detailed mapping with EDS chemical analyses is presented in Fig. 7 for all the composites. Regarding the surface microstructure and mapping of these composites, large aluminum particles (fresh scraps aluminum) were observed in the matrix. However, all the main reinforcements are distributed very homogenously. Fine TiAl and Nb_2Al intermetallic particles are homogeneously distributed and played a strong cohesion with the matrix. Much more efficient combined effect of two intermetallic particles was observed on the chemical diffusion bonding at the interfaces between the matrix and the reinforcements. As seen from the pictures (Fig. 7) taken from the composites, Nb_2Al intermetallic particles show a durable chemical diffusion bonding that contribute to the toughening mechanism. In fact, the strength values should be higher if the recycled aluminum scraps could be finer in size. In the frame of this research work, these values were obtained under laboratory conditions that provide a useful tool for industrial applications.

Conclusion

Innovative aluminum (Al 356 + A7075) matrix composites (AMCs) were designed from recycled fresh scrap-chips supplied by the Brazilian aeronautic company. The composites were manufactured through combined through the method of powder metallurgy followed by "sintering + forging." In this work, four different composites were designed using a low cost manufacturing process for aeronautical engineering applications. As the primary reinforcements, fine intermetallic powders Nb_2Al and TiAl were used. Also fine ZrO_2 were used and as the secondary reinforcements. In general, the static and incremental cyclic compression tests and low velocity impact tests give promising results. For the next stage of this joint research project, operational parameters for the static and cyclic tests and time-dependent parameters for low cycle fatigue tests should be taken into account to optimize the composition with the reinforcement levels.

Acknowledgments The authors thank the technical facilities of the UNICAMP/FEM, Professor R CARAM's lab. Mechanical tests (Zwick test device) and microstructural analyses have been carried out at ISAE-SUPMECA-Paris.

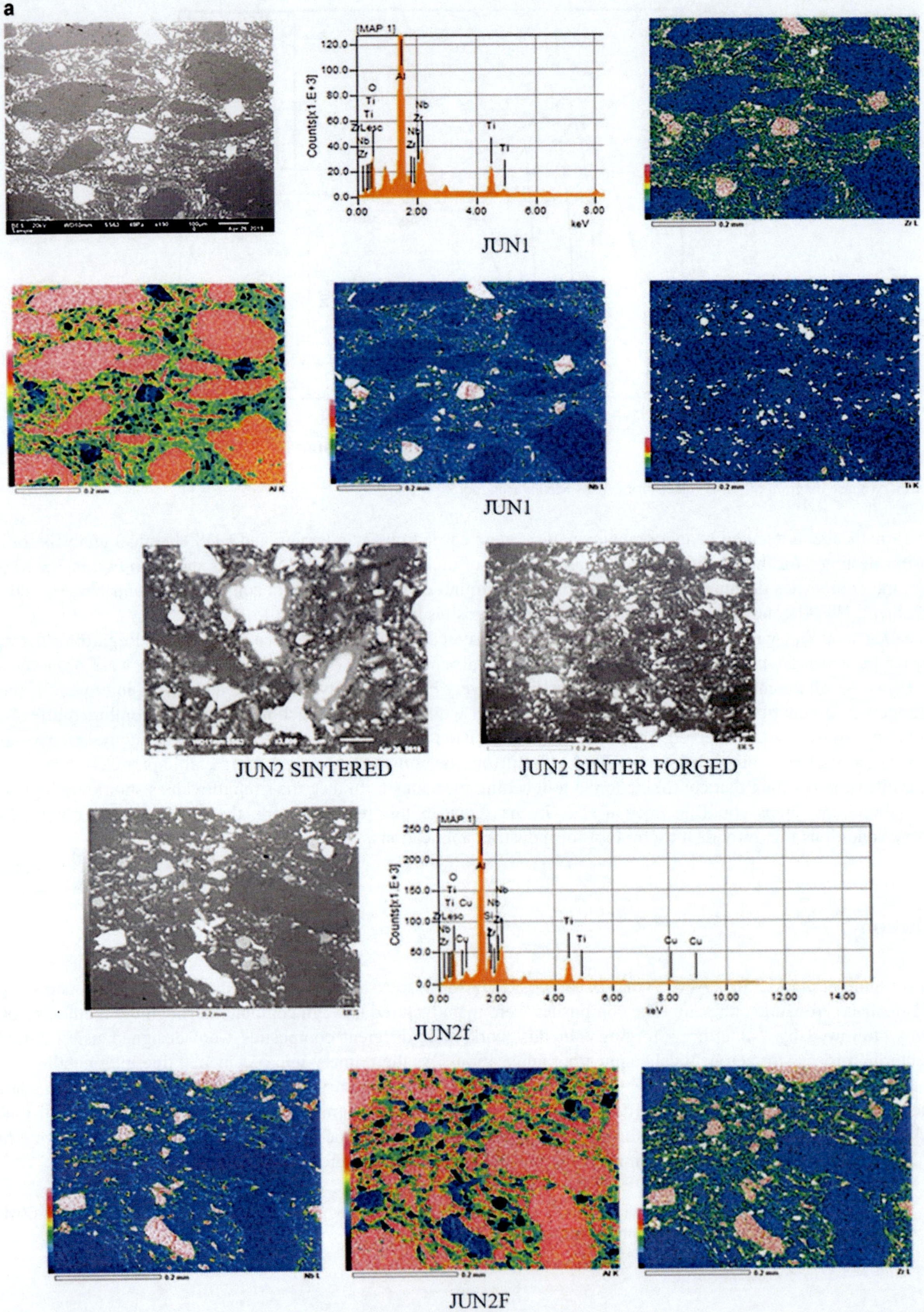

Fig. 7 Mapping and EDS analyses for the comparative study for the microstructure of the composite (**a**) JUN I and JUN II, (**b**) JUN III and JUN IV

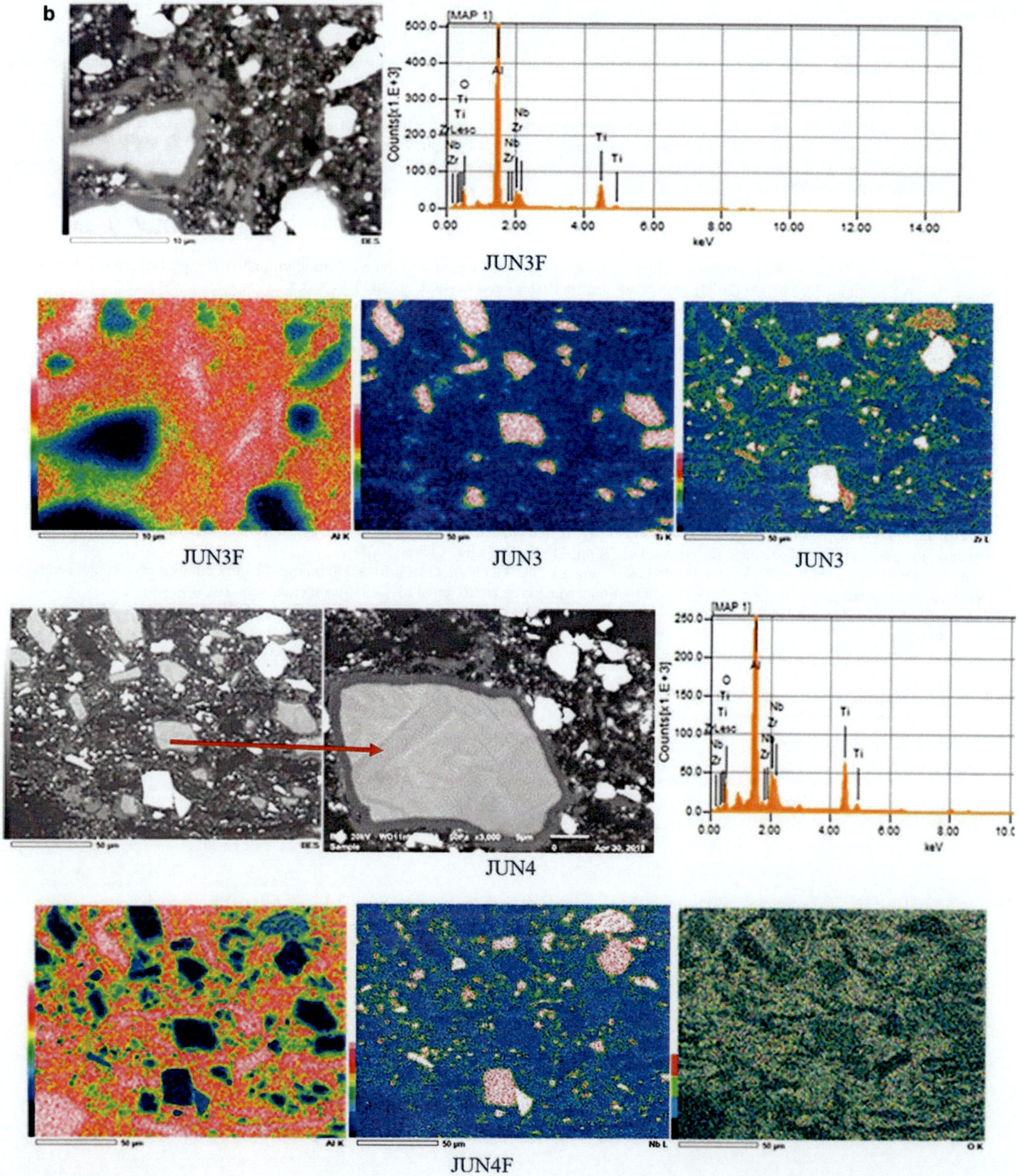

Fig. 7 (continued)

References

1. Gatamorta, F., Enginsoy, H.M., Bayraktar, E., Miskioglu, I., Katundi, D.: Design of recycled alumix-123 based composites reinforced with γ-Al_2O_3 through combined method; sinter + forging. In: Mechanics of Composite, Hybrid and Multifunctional Materials, vol. 6, pp. 9–17. Springer, Cham (2021)
2. Ravi Kumar, K., Pridhar, T., Sree Balaji, V.S.: Mechanical properties and characterization of zirconium oxide (ZrO_2) and coconut shell ash (CSA) reinforced aluminium (Al 6082) matrix hybrid composite. J. Alloys Compd. **765**, 171e179 (2018)
3. Bewlay, B.P., Jackson, M.R., Zhao, J.-C., Subramanian, P.R.: A review of very-high-temperature Nb-silicide–based composites. Metall. Mater. Trans.-A. **34A**, 2043–2555 (2003)
4. Enginsoy, H., Gatamorta, F., Bayraktar, E., Robert, M., Miskioglu, I.: Experimental and numerical study of Al-Nb_2Al composites via associated procedure of powder metallurgy and thixoforming. JCOMB, Compos. B.: Eng. **162**, 397–410 (2019). https://doi.org/10.1016/j.compositesb.2018.12.138
5. Zhang, S., Chen, Z., Wei, P.: Wear properties of graphene/zirconia biphase nano-reinforced aluminium matrix composites prepared by SLM. Mater. Today Commun. **30**, 103009 (2022). https://doi.org/10.1016/j.mtcomm.2021.103009
6. John, J.H.: Niobium alloys and high temperature applications. In: Niobium Science & Technology: Proceedings of the International Symposium Niobium 2001, Orlando, Florida, USA (2001)
7. Kurşun, A., Bayraktar, E., Enginsoy, H.M.: Experimental and numerical study of alumina reinforced aluminium matrix composites: processing, microstructural aspects and properties. Compos. Part B. **90**, 302–314 (2016)
8. Chawla, N., Sidhu, R.S., Ganesh, V.V.: Three-dimensional visualization and microstructure-based modeling of deformation in particle-reinforced composites. Acta Mater. **54**, 1541–1548 (2006)
9. Xu, Q., Qu, S.: Irreversible deformation of metal matrix composites: a study via the mechanism-based cohesive zone model. Mech. Mater. **89**, 72–84 (2015)
10. Breslin, M.C., Ringnalda, J., Xu, L., Fuller, M., Seeger, J., Daehn, G.S., Otani, T., Fraser, H.L.: Processing, microstructure, and properties of co-continuous alumina-aluminum composites. Mater. Sci. Eng. A. **195**, 113–119 (1995)
11. Mihlyuzova, L., Enginsoy, H.M., Slavov, S., Dontchev, D., Bayraktar, E.: Experimental and finite element study of recycled aluminium (AA7075) matrix composites reinforced of TiC/MoS_2/γ-Al_2O_3Fibre/Nb_2Al. In: Mechanics of Composite, Hybrid and Multifunctional Materials, Fracture, Fatigue, Failure and Damage Evolution, vol. 3, pp. 41–47. Springer, Cham (2022)
12. Sonia Ezeddini, D., Katundi, E.B., Miskioglu, I.: Laser cutting of the TiN + Al_2O_3 reinforced aluminium matrix composites through semisolid sintering. In: Mechanics of Composite and Multi-functional Materials, vol. 6, pp. 115–129. Springer, Cham (2017)
13. Zebarjad, S.M., Sajjadi, S.A.: Microstructure evaluation of Al-Al2O3 composite produced by mechanical alloying method. Mater. Des. **27**, 684–688 (2006)
14. Zebarjad, S.M., Sajjadi, S.A.: Dependency of physical and mechanical properties of mechanical alloyed Al-Al2O3 composite on milling time. Mater. Des. **28**, 2113–2120 (2007)
15. Fang, Z.Z.: Sintering of Advanced Materials. Woodhead Publishing, Cambridge (2010). ISBN 9781845695620
16. ASM Hand Book Binary Phase Diagrams. ASM International, USA (1979)

Dynamic Piezoresistive Behaviour of Composite Materials: Experimental Testing and Analytical Modelling

Mattia Utzeri, Attilio Lattanzi, Shanmugam Kumar, and Marco Sasso

Abstract Nowadays, additive manufacturing technologies allow coupling peculiar material properties with complex shapes to obtain cellular materials capable of exhibiting advanced multi-functionalities. Among them, self-sensing materials are increasingly valuable for applications where structural integrity monitoring is needed without external measurement instruments. This study exploits the piezoresistive properties of composite materials coupled with their own 3D-printed shapes. Therefore, understanding and modelling piezoresistive behaviour is getting a need. The piezoresistive behaviour of 3D printed composite material has been investigated under quasi-static and dynamic compression loadings. An innovative split Hopkinson bar set-up is introduced in order to measure the change in electrical resistance of composite material during the high strain rate compression. The strain rate and temperature effects on the material's piezoresistivity behaviour are discussed. Based on experimental evidence, a strain rate-dependent parameter is introduced into piezoresistivity analytical theory. The analytical findings are compared with the experimental ones.

Keywords Piezoresistivity · Dynamics · Composite materials · Self-sensing · Smart materials

Introduction

Self-sensing materials belong to smart materials, and they are materials able to monitor their properties by their own self, that is, without any measuring instrument applied to them. Among self-sensing materials, piezoresistive materials are extremely interesting materials because piezoresistive behaviour enables monitoring mechanical performance such as stress, strain, and damage evolution through changes in electrical resistance. Piezoresistive materials are deeply studied in literature, but they are usually referred to as natural materials [1]. Piezoresistive materials in nature exhibit giant piezoresistivity as silver, nickel, or carbon-based materials, but they cannot withstand high deformation. This disadvantage makes it impossible to create a self-sensing structure, and hence they are usually used as measurement instruments, that is, strain gauge base materials.

Additive manufacturing coupled with conductive dopant material enables the manufacturing of complex conductive composite cellular materials unthinkable in the past. Presently, it is possible to manufacture a self-sensing structure based on self-sensing composite materials. Consequently, additively manufactured smart materials are getting more and more important in cutting-edge applications. For example, the biomedical sector is focusing on smart materials due to the real need to monitor an implant without measuring instruments. Smart materials are also important in aerospace, automotive, security, defence, or industrial applications [1–3].

In literature, Kumar et al. tested the piezoresistivity of several additively manufactured metamaterials. They investigated the piezoresistive behaviour and the achievable gauge factor of several FFF composite materials, pointing out the potentiality of strong honeycombs and lattice structures [4, 5]. The polymer becomes piezoresistive once conductive dopant materials like carbon nanotubes or graphene are dispersed within it. Therefore, high-performance thermoplastic filaments have extreme potential in applications where the structural, thermal, and toxicity properties are the must-have properties. All these advantages coupled with piezoresistive properties bring FFF composite materials as highly valuable structural self-sensing materials [6, 7].

M. Utzeri (✉) · A. Lattanzi · M. Sasso
Department of Industrial Engineering and Mathematical Sciences, Polytechnique University of Marche, Ancona, Italy
e-mail: m.utzeri@pm.univpm.it

S. Kumar
James Watt School of Engineering, University of Glasgow, Glasgow, UK

F. Gardea et al. (eds.), *Mechanics of Composite, Hybrid and Multifunctional Materials, Volume 5*,
Conference Proceedings of the Society for Experimental Mechanics Series, https://doi.org/10.1007/978-3-031-50478-5_8

The main innovation in this study is evaluating the self-sensing properties of FFF-composite materials under dynamic loadings as none have done it before. The high strain rate piezoresistive material properties were investigated by coupling the traditional experimental mechanics technique (split Hopkinson pressure bar) with an electrical measurement instrument. Furthermore, a piezoresistive analytical model is discussed in order to figure out how the constitutive and geometrical nonlinearities affect the piezoresistive response.

Materials and Methods

Additive Manufacturing

The conductive material was printed via Fused Filament Fabrication (FFF). Cylindrical samples were printed through Apium printer with a nozzle of 0.4 mm. The cylinders were printed with 100% of infill so that the cylinder is solid. The cylinder was laid down on its side in order to align the loading directions with the beads' axis [8]. Printing parameters are based on our extensive testing in the laboratory. They ensure the best choice in strength, printability, and size accuracy. The cylinders have a diameter of 8 mm and are 8 mm long. The slenderness ratio of the cylindrical sample was chosen equal to 1 to ensure the dynamic equilibrium during the split Hopkinson bar tests [9].

Analytical Modelling

It is well established in solid-state physics that the electrical resistivity is a 3 × 3 tensor [10]. Quite all conductive materials exhibit diagonal resistivity tensors. The nonlinear piezoresistive effects change the tensor components' values, increasing or reducing the overall conductor electrical resistivity. Therefore, the nonlinear resistivity tensor is usually expressed as follows:

$$\overline{\overline{\rho}} = \begin{bmatrix} {\rho_1}^0 & 0 & 0 \\ 0 & {\rho_2}^0 & 0 \\ 0 & 0 & {\rho_3}^0 \end{bmatrix} + \begin{bmatrix} \Delta\rho_{11} & \Delta\rho_{12} & \Delta\rho_{13} \\ \Delta\rho_{21} & \Delta\rho_{22} & \Delta\rho_{23} \\ \Delta\rho_{31} & \Delta\rho_{32} & \Delta\rho_{33} \end{bmatrix}$$

where ${\rho_{ij}}^0$ is the resistivity at zero loading and $\Delta\rho_{ij}$ is the change in electrical resistivity, that is, $\Delta\rho_{ij} = \rho_{ij} - {\rho_{ij}}^0$. The tensor is re-written dividing all terms by the electrical resistivity at zero loading, so the electrical resistivity tensor takes the following form:

$$\overline{\overline{\rho}} = \left[\overline{\overline{I}} + \overline{\overline{r}}\right]\rho^0 = \begin{bmatrix} 1 & 0 & 0 \\ 0 & 1 & 0 \\ 0 & 0 & 1 \end{bmatrix}\rho^0 + \begin{bmatrix} r_{11} & r_{12} & r_{13} \\ r_{21} & r_{22} & r_{23} \\ r_{31} & r_{32} & r_{33} \end{bmatrix}\rho^0$$

where $r_{ij} = \frac{\rho_{ij} - {\rho_{ij}}^0}{{\rho_{ij}}^0}$ and they coincide with the differentiation of electrical resistivity, that is, $r_{ij} = \frac{d\rho_{ij}}{{\rho_{ij}}^0}$. In this study, the tensor $\overline{\overline{r}}$ is called the piezoresistance tensor. Thanks to this visualization, it is possible to figure out anisotropy phenomena due to external loading. In order to model the piezoresistivity effects, a relationship that connects the change in electrical resistivity with the external loadings needs to be introduced. In literature, physics dictates a strong connection between stress and piezoresistance [10]. The relationship between stress components and piezoresistance tensor takes the following form:

$$\overline{r} = \overline{\overline{\Pi}}\,\overline{\sigma}$$

where $\overline{\overline{\Pi}}$ is the piezoresistive stress matrix, and it connects stress to piezoresistance as well as the constitutive matrix in solids does. The tensor results easier to be visualized and managed once completely developed, namely

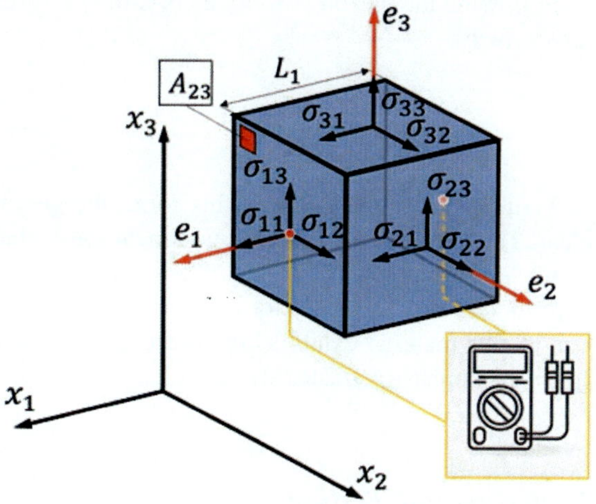

Fig. 1 Sketch of piezoresistive tensor on elementary conductor cube

$$\begin{bmatrix} r_{11} \\ r_{22} \\ r_{33} \\ r_{23} \\ r_{31} \\ r_{12} \end{bmatrix} = \begin{bmatrix} \Pi_{11} & \Pi_{12} & \Pi_{13} & \Pi_{14} & \Pi_{15} & \Pi_{16} \\ \Pi_{21} & \Pi_{22} & \Pi_{23} & \Pi_{24} & \Pi_{25} & \Pi_{26} \\ \Pi_{31} & \Pi_{32} & \Pi_{33} & \Pi_{34} & \Pi_{35} & \Pi_{36} \\ \Pi_{41} & \Pi_{42} & \Pi_{43} & \Pi_{44} & \Pi_{45} & \Pi_{46} \\ \Pi_{51} & \Pi_{52} & \Pi_{53} & \Pi_{54} & \Pi_{55} & \Pi_{56} \\ \Pi_{61} & \Pi_{62} & \Pi_{63} & \Pi_{64} & \Pi_{65} & \Pi_{66} \end{bmatrix} \begin{bmatrix} \sigma_{11} \\ \sigma_{22} \\ \sigma_{33} \\ \sigma_{23} \\ \sigma_{31} \\ \sigma_{12} \end{bmatrix}$$

where the $\Pi_{i\,j}$ is the piezoresistive stress components. The piezoresistive stress matrix can be manipulated to ensure also a relationship with strain [10, 11]. In the linear elastic stage, the piezoresistive strain matrix is defined according to the constitutive material matrix, namely

$$\overline{r} = \overline{\overline{\Pi}}\,\overline{\sigma} = \overline{\overline{\Pi}}\,\overline{\overline{C}}\,\overline{\varepsilon} = \overline{\overline{\Psi}}\,\overline{\varepsilon}$$

where the Ψ is the piezoresistive strain matrix and C is the stiffness matrix.

In this study, the piezoresistive behaviour is investigated under uniaxial compression test so the nonlinear resistivity is evaluated along the first principal direction as shown in Fig. 1. The electrical resistance is dictated by Ohm's law; hence, the electrical resistance along the first direction becomes

$$R_{11} = \rho_{11}\frac{L_1}{A_{23}} = \rho^0(1 + r_{11})\frac{L_1}{A_{23}}$$

where the L_1 is the conductor length and A_{23} is the cross-sectional area. To measure R_{11}, the experimental test must be set up with a multimeter along the first principal direction, that is, the pin must be connected to the upper and lower sides of the metamaterial. The length L_1 measures the long distance between the multimeter pins. A_{23} is the area where the pins touch as shown in Fig. 1. The change in electrical resistance along the first principal direction is written following the geometrical relationship [11]. Then, Eq. 1 is differentiated to provide information about how the change in resistance relates in geometrical or electrical features

$$\frac{dR_{11}}{R_0} = \frac{d\rho_{11}}{d\rho_0} + \varepsilon_1 - \varepsilon_2 - \varepsilon_3 = r_{11} + \varepsilon_1 - \varepsilon_2 - \varepsilon_3$$

Under monoaxial stress state, a conductive isotropic material exhibits the change in electrical resistance as follows:

$$\frac{dR_{11}}{R_0} = r_{11} + \varepsilon_1(1 + 2\nu)$$

Following the piezoresistivity theory, the piezoresistivity and gauge factor along the first principal direction can be defined, respectively.

$$PF_{11} = \frac{\frac{d\rho_{11}}{d\rho_0}}{\varepsilon_1} = \frac{r_{11}}{\varepsilon_1} \quad GF_{11} = \frac{\frac{dR_{11}}{dR_0}}{\varepsilon_1} = \mathrm{PF}_{11} + 1 + 2\nu$$

Writing the gauge factor in this form, the geometrical and constitutive contributes on piezoresistive properties are now clear. The gauge factor becomes nonlinear only when the conductors exhibit piezoresistivity ($\mathrm{PF}_{11} \neq 0$). Otherwise, the gauge factor is constant and Poisson's governs the phenomenon. When the PF factor is extremely higher then geometrical contributes, the gauge factor can be considered fully nonlinear and completely material dependent. In this study, the conductive material exhibits high piezoresistive behaviour, so the gauge factor is considered only piezoresistivity dependent (giant piezoresistive material).

Experimental Testing

The piezoresistive material properties were investigated by coupling the traditional experimental mechanics technique with an electrical measurement instrument. Both quasi-static and dynamic compression tests were carried out in the laboratory of the Polytechnique University of Marche. Quasi-static compression tests were carried out by Zwick/Roell©Z050 [12]. Dynamic compression tests were carried out by direct split Hopkinson bar (SHB) [8, 9]. The SHB system consists of three aligned steel bars (Pre-stressed, Input, and Output) of 18 mm in diameter and 3.0, 7.5, and 4.0 m long, respectively [8]. The mechanical test was coupled with an electrical measurement system that is able to monitor the change in electrical resistance between the compressive plates, that is, electrical resistance of compressed samples.

The piezoresistivity measure instrument is homemade, and it follows the traditional approach to identify an unknown electrical resistance in series circuit. Figure 2 shows how the electrical circuit works. In detail, the compressive plate was wired and the wire connected to the NI PCI-6120. The known electrical resistance R is wired and connected to the NI PCI-6120 and to the left side of the compressive plate. Thus, the R resistance is in series with the compressed sample. The electrical resistance of R should be comparable with the sample resistance at zero loading; in this case, R is 135,000 Ω. Otherwise, the piezoresistivity measurement becomes low sensitive. Both resistance R and sample were powered by the NI PCI-6120 with a constant tension, that is, 5 V. This set-up ensures measuring of the unknown electrical resistance, that is, the sample resistance R_U, through the simple relationship:

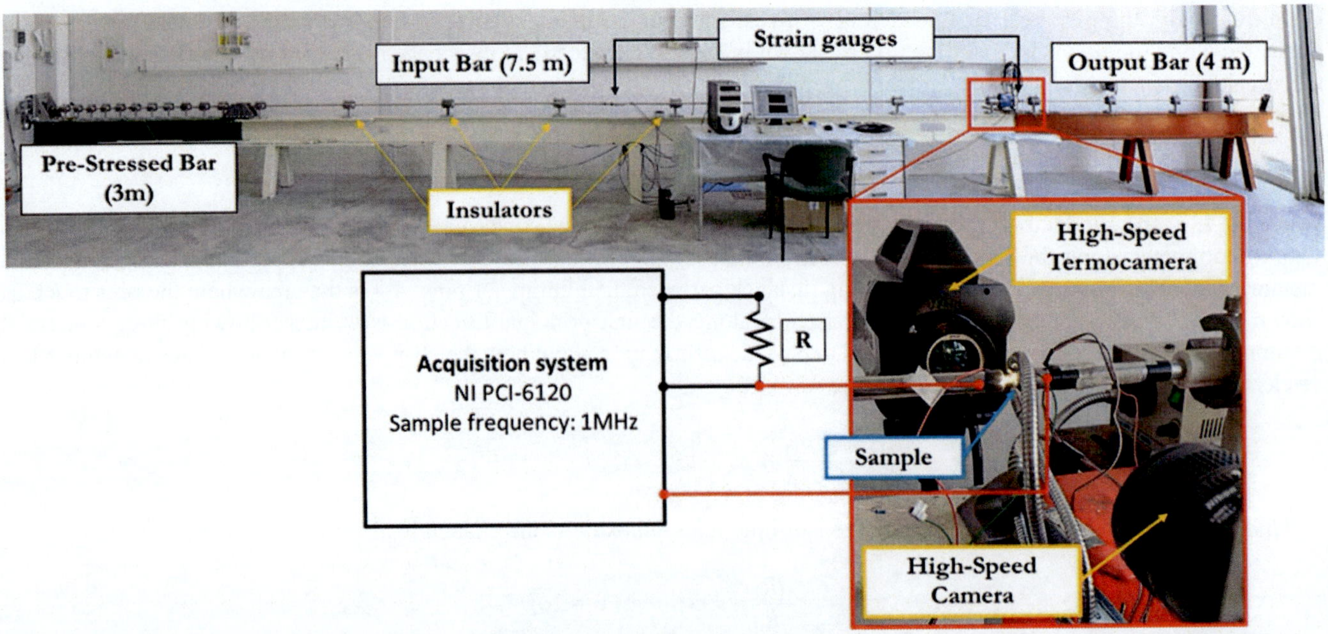

Fig. 2 Set-up of split Hopkinson pressure bar for high strain rate piezoresistivity measurement

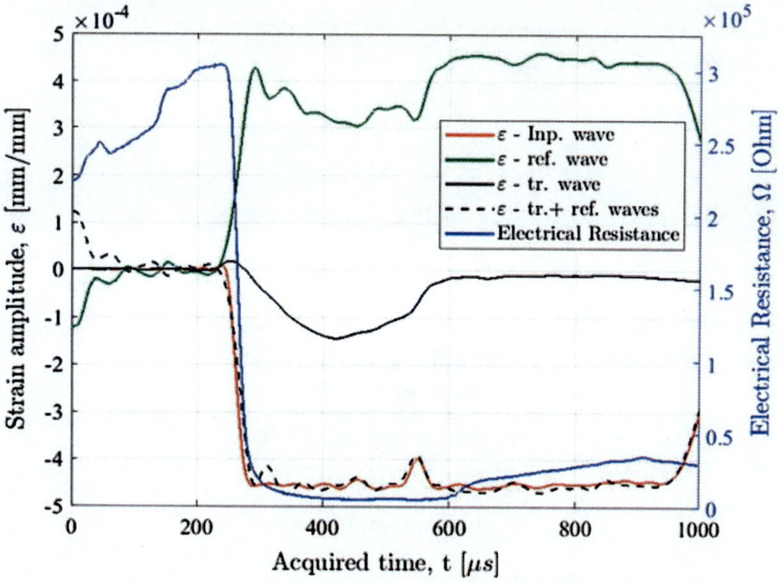

Fig. 3 Strain waves signals into input and output bars coupled with electrical resistance signal

$$R_U = \frac{\Delta V_U}{\Delta V_R} R$$

where ΔV_U is the potential measured by the instrument between the sample edges. ΔV_R is the potential measured between the known electrical resistance R. The measure is strongly dependent on the environment. The compressive plates must be insulated from the other components. Under quasi-static loading, insulators were placed under the compressive plate. Instead, in SHPB the insulator had not been placed along the bar. The bars were completely insulated from the structure through cylindrical support in PTFE, as shown in Fig. 2. The resistance of the sample was measured with a multimeter before the test to ensure the measure's effectiveness. Comparing both results is possible to figure out the effective sample insulation. The measurement set-up was used for both loading conditions. In this way, the high strain rate piezoresistive material properties are directly compared with the quasi-static ones. The compression tests were recorded by a high-speed camera (Fast-cam 150 kHz) and by a high-speed thermocamera (Flir 8 kHz). Cameras, strain gauges, and voltages were acquired by NI PCI-6120 at 1 MHz. The signals are triggered on input strain waves signal so the signals' acquisition starts at the same time and they are considered synchronized.

The strain signals are superposed in Fig. 3. The transmitted and reflected waves are shifted in time based on the strain gauge position as well as the electrical resistance history. The analytical shifting procedure highlights the dynamic equilibrium achieved during the compression test. The electrical resistance increases at the beginning because the pre-stressed bar is stretched so the compressive pre-loaded is slightly lost. Once the normal stress waves travel along the input bar, the electrical resistance is constant. The electrical resistance decreases only after the compressive wave gets the sample. In fact, the electrical resistance signal is coherently synchronized with the stress signal. The electrical resistance increases after the sample breaking as shown in Fig. 3 at about 600 μs. The electrical resistance history allows for figuring out when the failure occurred in turn. The stress curve drops quite instantly after the failure. However, is not easy to define the effective failure strain. The electrical resistance increase is a sign of fracture; hence, the failure strain is clearly defined.

Results and Discussions

The quasi-static and dynamic compression test results show a strain rate sensitivity of the FFF-printed material. The engineering stress-strain curve is shown in Fig. 4. The material exhibits a brittle behaviour under compression test in both cases. Young's modulus is not affected by strain rate, and it is about 3328 MPa. On the contrary, the peak stress jumps from about 100 MPa to 160 MPa at 500 1/s. The composite material does not lose ductility as the strain rate increases. The results confirm the good structural integrity achieved by the printing parameters [8].

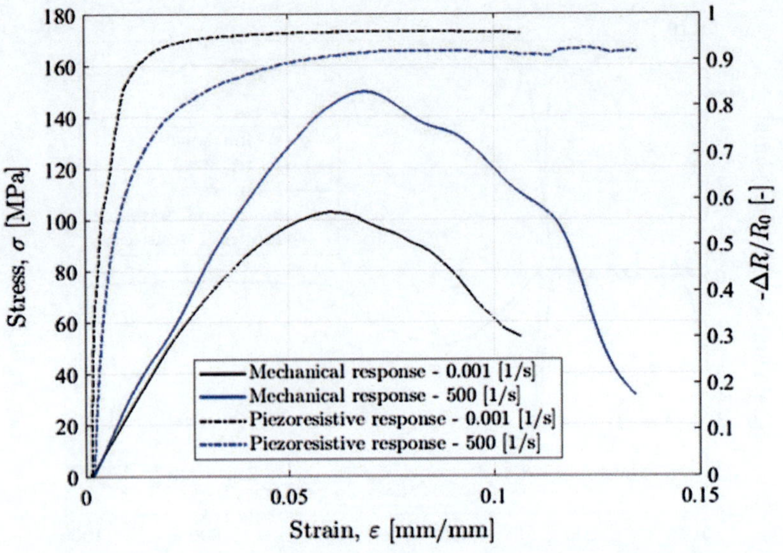

Fig. 4 Mechanical and piezoresistive behaviour of FFF-printed material under quasi-static and dynamic loading conditions

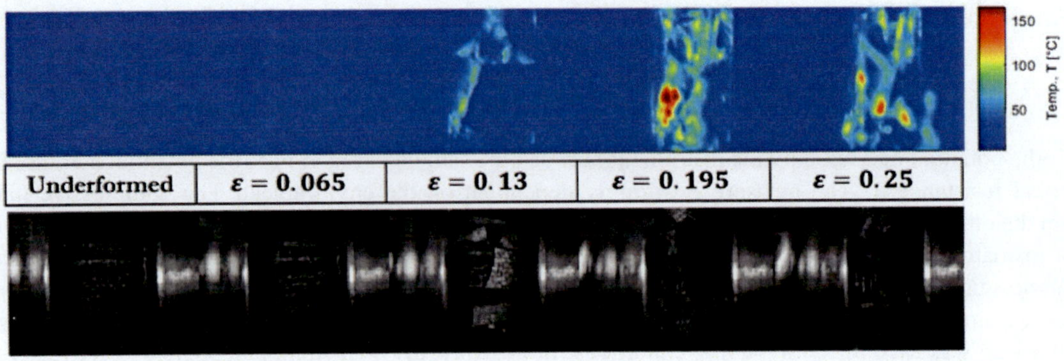

Fig. 5 High-speed thermocamera results versus deformated shape of FFF-printed sample in high strain rate compression test

The piezoresistive response is affected by the strain rate as well. In fact, the conductive dopant materials dispersed into a composite material matrix get closer once the compression loading is applied. The conductive paths connect fillers inside the matrix to become shorter, and, thus the conductivity higher. Therefore, the microstructure governs the change in resistivity [3]. Figure 4 shows the electrical resistance increases very sharply in quasi-static conditions. Actually, the electrical resistance is contact dependent. The roughness of the sample affects the electrical resistance response. However, the piezoresistive response is associated with the sample once the electrical resistance overtakes the values measured at zero loading (R_0 =115,000 Ω). On the contrary, the dynamic piezoresistive behavior is smoother than quasi-static conditions, also showing a lower peak electrical resistance.

The high strain rate affects the sample temperature and, in turn, the electrical resistivity due to it being temperature-dependent. Therefore, the temperature should be measured during the high-speed compression test. The high-speed thermocamera shows the temperature distribution during the compression in Fig. 5.

The temperature is homogeneous up to the peak of stress. After that, the cylinder starts to fracture increasing the temperature in the fractured zone from 15 °C at 0.065 mm/mm to 58 °C at 0.13 mm/mm. The temperature can be considered ineffective before the peak load; thus, the smoother piezoresistive response is accredited to strain rate effect. After the peak of stress, sample fragmentation makes difficult to figure out how the temperature could affect the piezoresistive response.

The gauge factors and piezoresistive factors are investigated as well. Figure 6 shows the gauge factor is extremely higher than geometrical (GF ≅ 2); hence, the gauge factor is completely dependent on constitutive nonlinearities, that is, the GF is comparable to PF. The strain rate effect reduces the GF at lower strain. In fact, the change in electrical resistance is smoother in quasi-static loading than in the dynamic one.

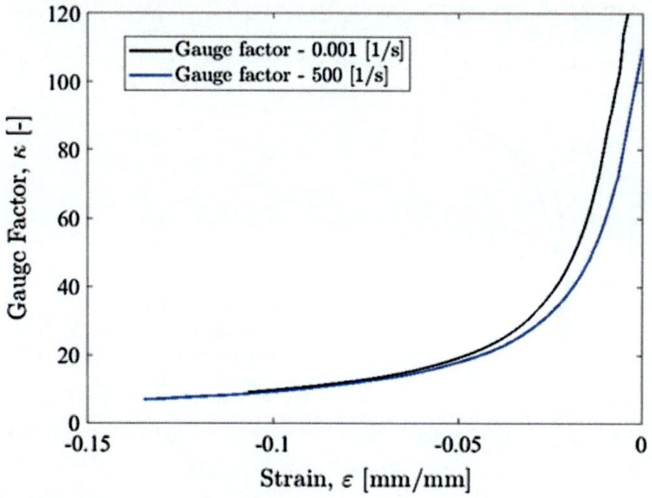

Fig. 6 Gauge factors obtained in quasi-static and dynamic condition

The gauge factors are superposed at high strain. However, the GF after the peak of stress should be addressed with different equations. The higher electrical resistance at zero loading prevents an evaluation of the change in electrical resistance at high strain. This means the gauge factors are not really superposed. Kumar et al. introduced the damage piezoresistivity factor to address that issue [4, 5].

In this study, the strain rate effect is investigated following the analytical theory. The fully nonlinear GF depends on the r_{11} because it is extremely high. Thus, the gauge factor is directly connected with the piezoresistance value, namely

$$\mathrm{GF}_{11} = k = \frac{r_{11}}{\varepsilon_1} + 1 + 2\nu \rightarrow k = \frac{r_{11}}{\varepsilon_1}$$

The GF is fitted through an analytical equation as its shape is mathematically simple to fit. The GF is fitted through a rational equation in which the strain rate sensitive parameter is introduced

$$r_{11} = \frac{\varepsilon_1}{C(\dot{\varepsilon}) - \varepsilon_1}$$

where $C(\dot{\varepsilon})$ is function of strain rate. It is 0.0038 under quasi-static condition and 0.0069 under dynamic condition. Once the piezoresistance value is defined, the change in electrical resistance is predicted. The electrical resistance becomes

$$\frac{dR_{11}}{R_0} = \frac{\varepsilon_1}{C(\dot{\varepsilon}) - \varepsilon_1}$$

The analytical predictions are compared with the experimental results in Fig. 7. The analytical predictions exhibit a consistent piezoresistive behaviour with experimental results. This means the strain rate-dependent parameter $C(\dot{\varepsilon})$ introduced in the analytical relationship enables the prediction of piezoresistivity taking into account the strain rate sensitivity effects.

Conclusion

The main innovation in this study is evaluating the self-sensing properties of FFF-composite materials under dynamic loadings. An innovative split Hopkinson bar set-up is introduced to measure the change in electrical resistance of composite material during high strain rate compression. The FFF-printed material exhibits strain rate sensitivity in mechanical and piezoresistive properties. The piezoresistive response is only accredited to the strain rate effect because the temperature effect is considered ineffective. A strain rate-dependent parameter is introduced into the theoretical description of the piezoresistivity phenomenon. The proposed analytical modelling consistently predicts the experimental findings.

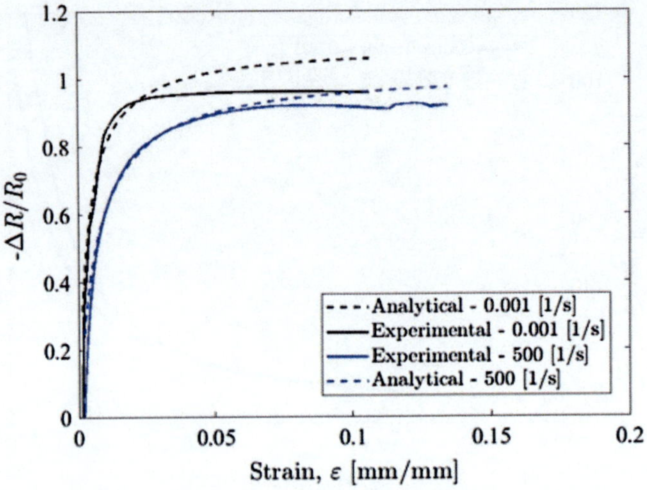

Fig. 7 Analytical versus experimental piezoresistive response of FFF-printed material

Acknowledgments Financed by the European Union-NextGenerationEU (National Sustainable Mobility Center CN00000023, Italian Ministry of University and Research Decree No.1033-17/06/2022, Spoke 11- Innovative Materials & Lightweighting), and National Recovery and Resilience Plan (NRRP), Mission 04 Component 2 Investment 1.5-NextGenerationEU, call for tender n. 3277 dated 30 December 2021. The opinions expressed are those of the authors only and should not be considered representative of the European Union or the European Commission's official position. Neither the European Union nor the European Commission can be held responsible for them.

References

1. Yang, C.F., Qin, L.J.: Graphical representation and explanation of the conductivity tensor of anisotropic media. Surv. Geophys. **41**(2), 249–281 (2020)
2. Xi, X., Chung, D.: Piezoresistivity and piezoelectricity discovered in aluminum, with relevance to structural self-sensing. Sensors Actuators A Phys. **289**, 144–156 (2019)
3. Chen, G., Lin, Y.: Stress–strain–electrical resistance effects and associated state equations for uniaxial rock compression. Int. J. Rock Mech. Min. Sci. **41**(2), 223–236 (2004)
4. Arif, M., Alhashmi, H., Varadarajan, K., Koo, J.H., Hart, A., Kumar, S.: Multifunctional performance of carbon nanotubes and graphene nanoplatelets reinforced PEEK composites enabled via FFF additive manufacturing. Compos. Part B. **184**, 107625 (2020)
5. Arif, M., Kumar, S., Varadarajan, K., Cantwell, W.: Performance of biocompatible PEEK processed by fused deposition additive manufacturing. Mater. Des. **146**, 249–259 (2018)
6. Katz, M.J.: Electrical conductivity in heavily doped n-type germanium: temperature and stress dependence. Phys. Rev. **140**(4A), A1323–A1344 (1965)
7. Utzeri, M., Morichelli, L., Lattanzi, A.: Influence of Core cellular structures on collapse mechanisms maps of sandwich beams. IOP Conf. Ser.: Mater. Sci. Eng. **1275**, 012009 (2023)
8. Utzeri, M., Farotti, E., Coccia, M., Mancini, E., Sasso, M.: High strain rate compression behaviour of 3d printed carbon-PA. J. Mater. Res. **36**(10), 2083–2093 (2021)
9. Utzeri, M., Bhagavatam, A., Mancini, E., Dinda, G., Sasso, M., Newaz, G.: Quasistatic and dynamic behavior of inconel 625 obtained by laser metal deposition: experimental characterization and constitutive modeling. J. Eng. Mater. Technol. **143**(4), 041007 (2021)
10. Ventura, I.A., Zhou, J., Lubineau, G.: Drastic modification of the piezoresistive behavior of polymer nanocomposites by using conductive polymer coatings. Compos. Sci. Technol. **117**, 342–350 (2015)
11. Chung, D.D.L.: A critical review of piezoresistivity and its application in electrical-resistance-based strain sensing. J. Mater. Sci. **55**(32), 15367–15396 (2020)
12. Rossi, M., Lattanzi, A., Cortese, L., Amodio, D.: An approximated computational method for fast stress reconstruction in large strain plasticity. Int. J. Numer. Methods Eng. **121**, 3048–3065 (2020)

Investigating the Fracture Resistance of Carbon Fiber-PEEK Composites Produced via Fused Filament Fabrication

Denizhan Yavas

Abstract The use of carbon fiber-reinforced polyetheretherketone (PEEK) produced through additive manufacturing is crucial due to its exceptional physical and mechanical characteristics, such as its high strength, stiffness, temperature resistance, chemical resistance, lightweight, and biocompatibility. This makes carbon fiber-reinforced PEEK a valuable material in various industries like aerospace, automotive, medical, and electronics. This study investigates the effect of short carbon fibers on the mode-I fracture behavior of PEEK composites made through additive manufacturing. The research employed Fused Filament Fabrication (FFF) to create both short carbon fiber-reinforced and non-reinforced PEEK as the baseline. A compact tension fracture test was conducted to measure the mode-I fracture toughness, and elastic-plastic fracture theory was used to calculate the inter-layer and cross-layer crack growth resistance. Unreinforced PEEK samples showed stable crack growth with significant plastic energy dissipation, whereas short fiber reinforcement resulted in unstable crack growth with subtle plastic energy dissipation. For the examined fiber volume fraction and utilized FFF parameters, the carbon fiber caused a significant decrease in the mode-I fracture toughness of 3D-printed PEEK. The inclusion of fiber reinforcement led to a reduction of approximately 50% in elastic fracture energy and about 80% in total fracture energy. This observation can be attributed to the high fiber volume percentage.

Keywords 3D printing · Fiber reinforcement · Composite materials · Fracture

Introduction

Carbon fiber-reinforced (CFR) polyetheretherketone (PEEK) composites have gained popularity in the aerospace and biomedical industries due to their remarkable physical and mechanical properties, such as high strength, stiffness, temperature resistance, chemical resistance, lightweight, and biocompatibility [1]. However, producing CFR-PEEK composites with complex geometries structures using traditional methods remains a challenge. 3D printing has emerged as a solution to these limitations, as it allows for the fabrication of CFRP composites with unique advantages [2, 3]. Among the various additive manufacturing (AM) techniques, such as Selective Laser Sintering (SLS), the Fused Filament Fabrication (FFF) method has become the most popular method for processing PEEK parts due to its cost-effectiveness, greater material flexibility, and high recyclability that leads to less waste. Despite these advantages, the FFF of PEEK is still a difficult process due to the intricate relationship between AM process parameters, microstructure, and the mechanical/fracture performance of the final product.

The current literature shows a certain degree of effort to understand the correlation between the FFF process parameters such as layer thickness, nozzle temperature, printing speed, and the resulting mechanical properties (i.e., tensile and flexural strength, ductility, and toughness) of additively manufactured PEEK parts [4–15]. However, there is a limited understanding of the fracture behavior of PEEK fabricated by FFF. The FFF of fiber-reinforced PEEK composites is an even more complicated process since it entails proper synthetization of fiber-reinforced PEEK filaments and strong physical and chemical compatibility between fibers and the PEEK matrix. Such an involved process necessitates a deeper physical understanding of the joint interaction between microstructure, chemistry, and micro-mechanical behavior of the fiber-matrix interface. A certain degree of effort has been made toward understanding the evolution of the mechanical and physical properties of fiber-reinforced PEEK composites with various FFF process parameters [16–28]. However, little is known about the complex correlation between the fracture behavior of the CFR-PEEK composites and the AM process parameters. In this study, we investigate the effect of short carbon fibers on the mode-I fracture behavior of PEEK composites made through additive

D. Yavas (✉)
Department of Mechanical Engineering, Rice University, Houston, TX, USA
e-mail: denizhan.yavas@rice.edu

F. Gardea et al. (eds.), *Mechanics of Composite, Hybrid and Multifunctional Materials, Volume 5*,
Conference Proceedings of the Society for Experimental Mechanics Series, https://doi.org/10.1007/978-3-031-50478-5_9

manufacturing. The research employed FFF to create both short carbon fiber-reinforced and non-reinforced PEEK as the baseline. A compact tension fracture test was conducted to measure the mode-I fracture toughness, and elastic-plastic fracture theory was used to calculate the effective fracture toughness and energy dissipation.

Compact Tension Specimen Fabrication

An FFF based 3D printer (Roboze Argo 500) was used to fabricate the compact tension (CT) fracture samples. This study uses PEEK and short carbon fiber (CF)-reinforced PEEK filaments with a nominal diameter of 1.75 mm. The volumetric percentage of the short carbon fibers is about 20%. All CT samples were printed with a 100% infill density and a layer thickness of 100 μm. The build orientation of XYZ (ASTM 52921 [29]) was selected to characterize the in-plane fracture toughness as shown in Fig. 1a. The raster orientation was varied by 90° between the adjacent layers to achieve a cross-ply [0°/90°] layup as shown in Fig. 1b, which provides an in-plane quasi-isotropic mechanical response. Figure 2a, b shows the

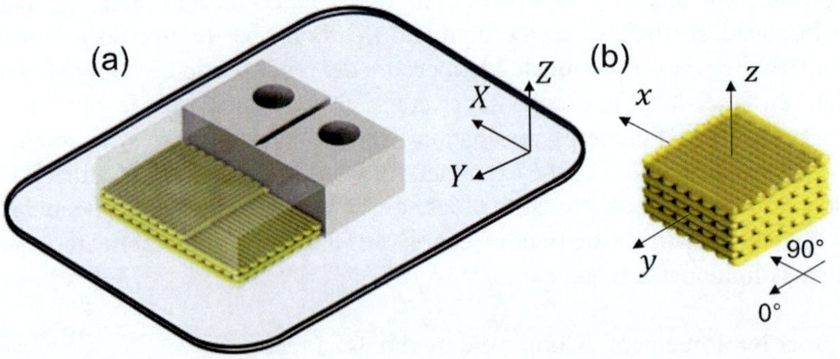

Fig. 1 (**a**) A depiction of the orientation and construction direction of the 3D-printed compact tension specimen in relation to the printing bed, and (**b**) the cross-ply infill pattern employed

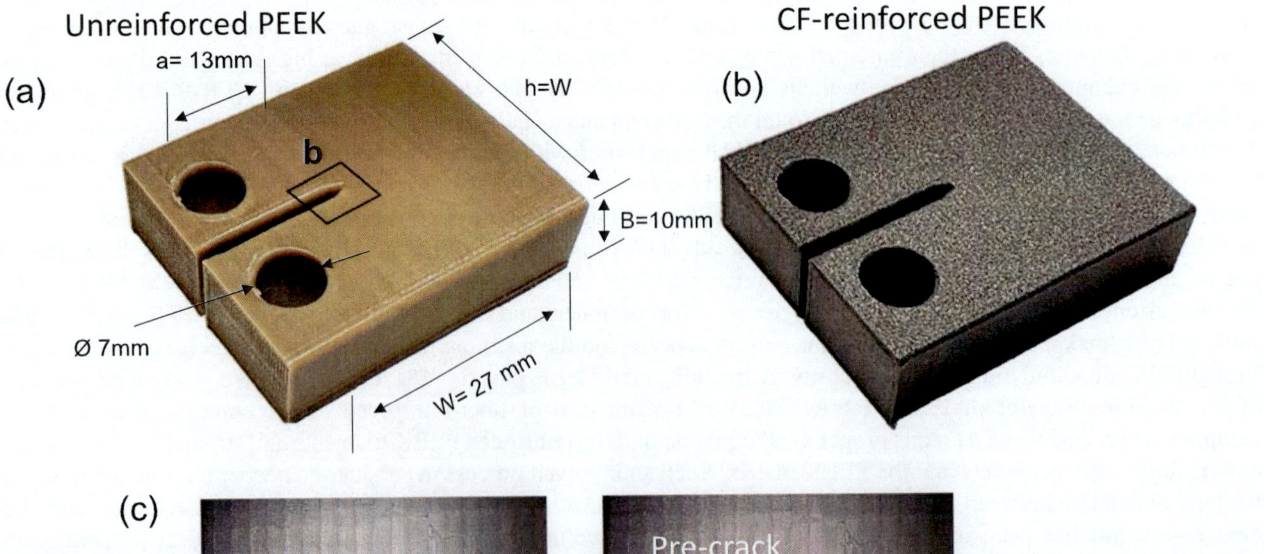

Fig. 2 Compact tension fracture samples fabricated by (**a**) unreinforced PEEK and (**b**) CF-reinforced PEEK. (**c**) Optical microscopy images showing the printed wedge before and after the crack sharpening procedure

geometry and dimension of the 3D-printed PEEK and CF-reinforced PEEK CT samples, respectively. The CT geometry was adapted from ASTM 5045-14 [30]. The pre-crack notch was created during the fabrication process, and then the notch tip was lightly struck using a fresh razor blade and soft mallet to achieve a sharp-crack tip through the width of the samples as depicted in Fig. 2c. The sample thickness B is 10 mm, which is sufficiently large to ensure the plane strain conditions. In addition, the initial length of the uncracked ligament (W-a) was adjusted to be long enough to enclose the entire plastic process zone in front of the crack tip.

Compact Tension Test Procedure

Compact tension (CT) tests were performed to measure the fracture toughness of the examined samples. The tests were conducted on an electromechanical universal testing frame, the Instron 50kN, as depicted in Fig. 3. A crosshead displacement rate of 2 mm/min was maintained for all the tests, and the force, crosshead displacement, and optical images of the samples were recorded. The monochrome camera used for capturing the images was a high-resolution FLIR Grasshopper 3, paired with an 85 mm lens, NIKKOR 24-85 mm f/2.8-4D. Images were acquired at a rate of 0.5 Hz and had a resolution of approximately 20 μm/pixel. To enable full-field strain measurement using the Digital Image Correlation (DIC) method, a random speckle pattern was applied to the surface of the samples using black and white spray paints, with an average size of 50 μm. The collected images were analyzed using Ncorr [31], an open-source software based on Matlab, to determine the full-field strain distribution across the hard and soft materials. A subject radius of 50 pixels was set in the DIC analysis to ensure that at least three speckle features were present within the selected window size. The full-field strain fields were used to identify the details of the growth nucleation and propagation.

The total mode-I fracture energy, J_0, was calculated using the principles of the elastoplastic fracture mechanics [32–36]. According to ASTM E1820 [32], J_0 is the confluence of the elastic strain energy dissipation J_{el} and plastic strain energy dissipation J_{pl}:

$$J_0 = J_{el} + J_{pl} \tag{1}$$

The elastic strain energy J_{el} is given by

$$J_{el} = \frac{K^2(1-\nu^2)}{E} \tag{2}$$

where E is the Young's modulus and ν is the Poisson's ratio of the examined material. The Young's modulus are $E = 3.0$ GPa and 3.5 GPa for the PEEK and CF-reinforced PEEK samples. K is the stress intensity factor which can be calculated for force P_i based on ASTM E1820 [32]:

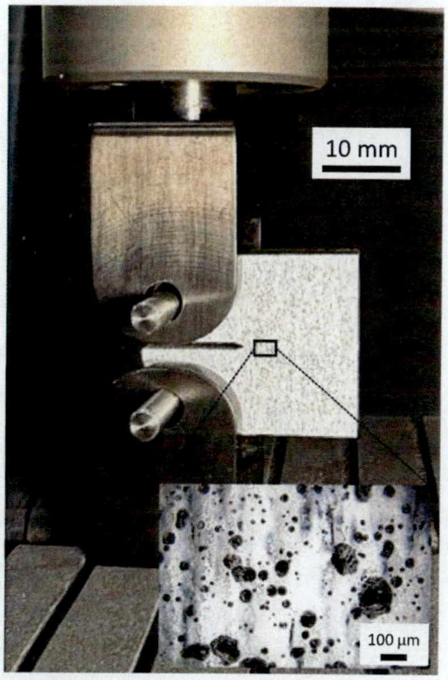

Fig. 3 Compact tension test setup. The inset image shows a close-up view of the sample surface depicting the random speckle pattern

$$K = \frac{P_i}{BW^{\frac{1}{2}}} f\left(\frac{a_i}{W}\right) \tag{3}$$

where B is the sample thickness, W is the sample width, and a_i is the crack length at force P_i. The function $f\left(\frac{a_i}{W}\right)$ is determined by the geometry of the compact tension sample. The critical stress intensity factor (i.e., mode-I fracture toughness) K_{IC} can be calculated from Eq. 3 by replacing P_i with critical fracture load P_Q.

The plastically dissipated energy J_{pl} is

$$J_{pl} = \frac{\eta_{pl} A_{pl}}{B b_o} \tag{4}$$

where A_{pl} is the area under the force-displacement curve and b_o is the uncracked ligament length. η_{pl} is a geometric factor and can be expressed in terms of the uncracked ligament length and sample width

$$\eta_{pl} = 2 + 0.522 b_o / W \tag{5}$$

Results

Figure 4a displays a representative set of force-displacement curves obtained from CT tests on unreinforced and carbon fiber (CF)-reinforced PEEK samples. Additionally, Fig. 4b, c provides snapshots of the opening strain fields measured by DIC at different stages of deformation marked in the force-displacement curves for the unreinforced and CF-reinforced PEEK samples, respectively. Initially, both samples deform linearly under loading, but the CF-reinforced sample has a higher initial stiffness due to the inclusion of fibers. This difference in stiffness can be attributed to the enhanced elastic modulus of the CF-reinforced sample compared to the unreinforced sample, considering they have the same pre-crack length. The force-displacement curves exhibit nonlinearity before reaching the maximum force level due to the establishment of a process zone ahead of the pre-crack tip before crack growth. At the peak force, as shown in Fig. 4b, c (panels B and F), the fracture process zone appears slightly bigger for the unreinforced PEEK sample than for the CF-reinforced PEEK sample. In the unreinforced PEEK sample, stable crack propagation commences at the peak force and proceeds through force decay. The opening strain presented in panels D and C of Fig. 4b shows the process zone size during the stable crack growth stage, which increases as the crack propagates. In contrast, the crack nucleates at a relatively smaller peak force in the CF-reinforced PEEK sample and propagates unstably during the sudden force drop. Panel F of Fig. 4c shows the fully developed fracture process zone prior to the unstable crack growth.

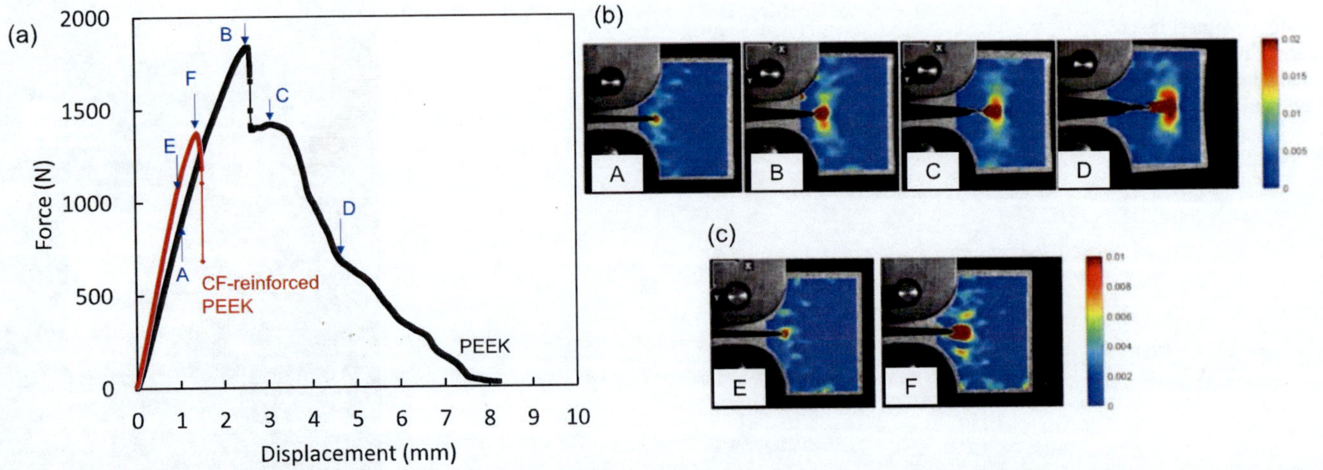

Fig. 4 (a) A representative set of force-displacement curves obtained from CT test of unreinforced and CF-reinforced PEEK. (b, c) The snapshots of the opening strain fields measured by DIC at the different stages of the deformation marked in the force-displacement curves for the unreinforced and CF-reinforced PEEK samples, respectively

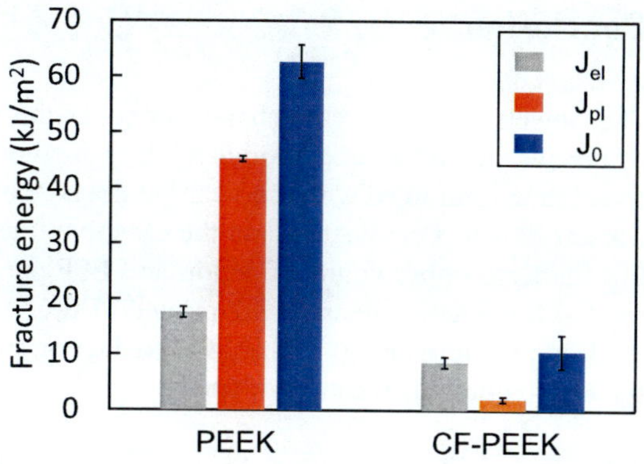

Fig. 5 The measured elastic, plastic, and total energy dissipation for the unreinforced and CF-reinforced PEEK samples. The error bars denote the standard deviation over the three experiments

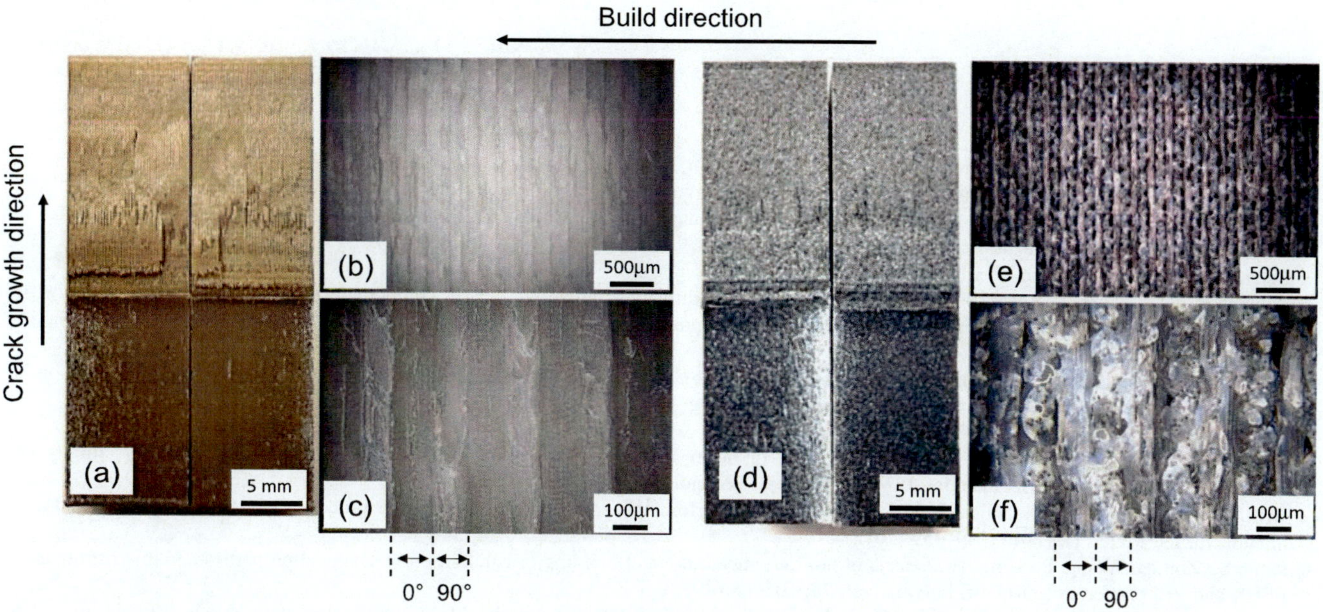

Fig. 6 Optical microscopy images of the fracture surfaces of (**a–c**) unreinforced PEEK and (**d–f**) CF-reinforced PEEK samples

Figure 5 presents the results of the measurements of elastic, plastic, and total energy dissipation for both unreinforced and CF-reinforced PEEK samples. The average measurements are reported, with error bars indicating the standard deviation across three experiments. The CF-reinforced PEEK samples exhibited about twofold lower fracture toughness in terms of elastic fracture energy dissipation when compared to the baseline PEEK. The difference became more significant in the case of plastic fracture energy dissipation, with a reduction of approximately 75% due to fiber reinforcement. This can be attributed to the transition of crack growth stability resulting from CF reinforcement. For the examined fiber volume fraction and FFF parameters, carbon fiber caused a significant decrease in the mode-I fracture toughness of 3D-printed PEEK. Additionally, the inclusion of fiber reinforcement resulted in a reduction of approximately 50% in elastic fracture energy and around 80% in total fracture energy. This observation can be explained by the high fiber volume percentage.

Figure 6 illustrates the fracture surfaces of unreinforced PEEK and CF-reinforced PEEK samples as observed through optical microscopy. The fracture surface of the unreinforced PEEK displays a rougher morphology compared to that of the CF-reinforced PEEK, resulting from the stretching and crazing of PEEK filaments. The micro-scale analysis of Fig. 6b, c reveals crazing marks on the surfaces of 90-degree oriented layers. In contrast, the fracture surface of the CF-reinforced PEEK appears smooth and glassy at the micro-scale as shown in Fig. 6e, f, as a result of brittle and unstable crack growth. It is worth noting that short carbon fibers are still embedded in the PEEK polymer, which may trigger micro-toughening mechanisms such as fiber pullout during the fracture process. These mechanisms have the potential to enhance the fracture toughness of the material.

Conclusion

The objective of this research was to study the impact of short carbon fibers on the mode-I fracture behavior of PEEK composites manufactured through additive methods. Fused Filament Fabrication was utilized to fabricate PEEK samples that were either reinforced with short carbon fibers or not reinforced. To measure the mode-I fracture toughness, a compact tension fracture test was conducted, and the effective fracture energy dissipated was calculated using elastic-plastic fracture theory. For the tested fiber volume fraction and FFF parameters, the addition of carbon fiber caused a significant reduction in the mode-I fracture toughness of 3D-printed PEEK. Additionally, the presence of fiber reinforcement resulted in a 50% reduction in elastic fracture energy and an 80% reduction in total fracture energy. This decrease in fracture energy can be attributed to the high-volume percentage of fibers.

Acknowledgments The author wishes to convey appreciation to Roboze company for their aid in sample creation and acknowledges the generous support received from the Center for Midstream Management and Science at Lamar University, under the leadership of Thomas Kalb. Special thanks are extended to Dr. Xuejun Fan for generously providing access to the testing equipment, and gratitude is expressed to undergraduate students Mr. Bradley Oliver and Mr. Lakota Jaton, as well as graduate student Mr. Luis Sosa, for their assistance in testing.

References

1. El Magri, A., El Mabrouk, K., Vaudreuil, S., Chibane, H., Touhami, M.E.: Optimization of printing parameters for improvement of mechanical and thermal performances of 3D printed poly (ether ether ketone) parts. J. Appl. Polym. Sci. **137**(37), 49087 (2020)
2. De Leon, A.C., Chen, Q., Palaganas, N.B., Palaganas, J.O., Manapat, J., Advincula, R.C.: High-performance polymer nanocomposites for additive manufacturing applications. React. Funct. Polym. **103**, 141–155 (2016)
3. Beland, S.: High-Performance Thermoplastic Resins and Their Composites. William Andrew (1990)
4. Rahman, K.M., Letcher, T., Reese, R.: Mechanical properties of additively manufactured PEEK components using fused filament fabrication. In: ASME International Mechanical Engineering Congress and Exposition, vol. 57359, p. V02AT02A009. American Society of Mechanical Engineers, New York (2015)
5. Xiaoyong, S., Liangcheng, C., Honglin, M., Peng, G., Zhanwei, B., Cheng, L.: Experimental analysis of high temperature PEEK materials on 3D printing test. In: 2017 9th International Conference on Measuring Technology and Mechatronics Automation (ICMTMA), pp. 13–16. IEEE (2017)
6. Yang, C., Tian, X., Li, D., Cao, Y., Zhao, F., Shi, C.: Influence of thermal processing conditions in 3D printing on the crystallinity and mechanical properties of PEEK material. J. Mater. Process. Technol. **248**, 1–7 (2017)
7. Arif, M.F., Kumar, S., Varadarajan, K.M., Cantwell, W.J.: Performance of biocompatible PEEK processed by fused deposition additive manufacturing. Mater. Des. **146**, 249–259 (2018)
8. Ding, S., Zou, B., Wang, P., Ding, H.: Effects of nozzle temperature and building orientation on mechanical properties and microstructure of PEEK and PEI printed by 3D-FDM. Polym. Test. **78**, 105948 (2019)
9. Davies, R., Shyng, Y.T., Wang, Y., Ghita, O.: Extrusion deposition of carbon nanotubes (CNT)/poly ether ether ketone (PEEK). In: ICCM International Conference on Composite Materials, pp. 1322–1328, July 2015
10. Wu, W., Geng, P., Li, G., Zhao, D., Zhang, H., Zhao, J.: Influence of layer thickness and raster angle on the mechanical properties of 3D-printed PEEK and a comparative mechanical study between PEEK and ABS. Materials. **8**(9), 5834–5846 (2015)
11. Liaw, C.Y., Tolbert, J.W., Chow, L.W., Guvendiren, M.: Interlayer bonding strength of 3D printed PEEK specimens. Soft Matter. **17**(18), 4775–4789 (2021)
12. Chen, W., Zhang, X., Tan, D., Xu, P., Yang, B., Shi, K., Zhu, B., Liu, Q., Lei, Y., Liu, S., Xue, L.: Improvement in mechanical properties of 3D-printed PEEK structure by nonsolvent vapor annealing. Macromol. Rapid Commun. **43**(7), 2100874 (2022)
13. Velu, R., Vaheed, N., Ramachandran, M.K., Raspall, F.: Experimental investigation of robotic 3D printing of high-performance thermoplastics (PEEK): a critical perspective to support automated fibre placement process. Int. J. Adv. Manuf. Technol. **108**(4), 1007–1025 (2020)
14. Wang, P., Zou, B., Xiao, H., Ding, S., Huang, C.: Effects of printing parameters of fused deposition modeling on mechanical properties, surface quality, and microstructure of PEEK. J. Mater. Process. Technol. **271**, 62–74 (2019)
15. Qin, Y., Ge, G., Yun, J., Tian, X., Liu, X., Han, J., Gao, S.: Enhanced impregnation behavior and interfacial bonding in CF/PEEK prepreg filaments for 3D printing application. J. Mater. Res. Technol. **20**, 4608–4623 (2022)
16. Wang, P., Zou, B., Ding, S., Huang, C., Shi, Z., Ma, Y., Yao, P.: Preparation of short CF/GF reinforced PEEK composite filaments and their comprehensive properties evaluation for FDM-3D printing. Compos. Part B. **198**, 108175 (2020)
17. Sommacal, S., Matschinski, A., Drechsler, K., Compston, P.: Characterisation of void and fiber distribution in 3D printed carbon-fiber/PEEK using X-ray computed tomography. Compos. A: Appl. Sci. Manuf. **149**, 106487 (2021)
18. Peng, W.A.N.G., Bin, Z.O.U., Shouling, D.I.N.G., Lei, L.I., Huang, C.: Effects of FDM-3D printing parameters on mechanical properties and microstructure of CF/PEEK and GF/PEEK. Chin. J. Aeronaut. **34**(9), 236–246 (2021)
19. Stepashkin, A.A., Chukov, D.I., Senatov, F.S., Salimon, A.I., Korsunsky, A.M., Kaloshkin, S.D.: 3D-printed PEEK-carbon fiber (CF) composites: structure and thermal properties. Compos. Sci. Technol. **164**, 319–326 (2018)
20. Yang, D., Cao, Y., Zhang, Z., Yin, Y., Li, D.: Effects of crystallinity control on mechanical properties of 3D-printed short-carbon-fiber-reinforced polyether ether ketone composites. Polym. Test. **97**, 107149 (2021)

21. El Magri, A., Vanaei, S., Vaudreuil, S.: An overview on the influence of process parameters through the characteristic of 3D-printed PEEK and PEI parts. High Perform. Polym. **33**(8), 862–880 (2021)
22. Luo, M., Tian, X., Shang, J., Zhu, W., Li, D., Qin, Y.: Impregnation and interlayer bonding behaviours of 3D-printed continuous carbon-fiber-reinforced poly-ether-ether-ketone composites. Compos. A: Appl. Sci. Manuf. **121**, 130–138 (2019)
23. Jiang, H., Aihemaiti, P., Aiyiti, W., Kasimu, A.: Study of the compression behaviours of 3D-printed PEEK/CFR-PEEK sandwich composite structures. Virtual Phys. Prototyp. **17**(2), 138–155 (2022)
24. Rodzeń, K., Harkin-Jones, E., Wegrzyn, M., Sharma, P.K., Zhigunov, A.: Improvement of the layer-layer adhesion in FFF 3D printed PEEK/carbon fibre composites. Compos. A: Appl. Sci. Manuf. **149**, 106532 (2021)
25. Luo, M., Tian, X., Shang, J., Yun, J., Zhu, W., Li, D., Qin, Y.: Bi-scale interfacial bond behaviors of CCF/PEEK composites by plasma-laser cooperatively assisted 3D printing process. Compos. A: Appl. Sci. Manuf. **131**, 105812 (2020)
26. Wang, P., Zhang, H., Cao, Y., Pan, A., Wu, W.: Additively manufactured short carbon fiber reinforced polyetheretherketone by coating polyetherimide at the interface using fused filament fabrication. J. Appl. Polym. Sci. **139**(26), e52438 (2022)
27. Chen, Y., Shan, Z., Yang, X., Song, Y., Zou, A.: Preparation of CCF/PEEK filaments together with property evaluation for additive manufacturing. Compos. Struct. **281**, 114975 (2022)
28. Prechtel, A., Reymus, M., Edelhoff, D., Hickel, R., Stawarczyk, B.: Comparison of various 3D printed and milled PAEK materials: effect of printing direction and artificial aging on Martens parameters. Dent. Mater. **36**(2), 197–209 (2020)
29. ASTM International: ISO/ASTM52921-13(2019) Standard Terminology for Additive Manufacturing—Coordinate Systems and Test Methodologies. ASTM International, West Conshohocken (2019). https://doi.org/10.1520/ISOASTM52921-13R19
30. ASTM International: D5045-14 Standard Test Methods for Plane-Strain Fracture Toughness and Strain Energy Release Rate of Plastic Materials. ASTM International, West Conshohocken (2014). https://doi.org/10.1520/D5045-14
31. Blaber, J., Adair, B., Antoniou, A.: Ncorr: open-source 2D digital image correlation matlab software. Exp. Mech. **55**(6), 1105–1122 (2015)
32. ASTM International: E1820–20 Standard Test Method for Measurement of Fracture Toughness. ASTM International, West Conshohocken (2020). https://doi.org/10.1520/E1820-20
33. Yavas, D., Zhang, Z., Liu, Q., Wu, D.: Interlaminar shear behavior of continuous and short carbon fiber reinforced polymer composites fabricated by additive manufacturing. Compos. Part B. **204**, 108460 (2021)
34. Yavas, D., Zhang, Z., Liu, Q., Wu, D.: Fracture behavior of 3D printed carbon fiber-reinforced polymer composites. Compos. Sci. Technol. **208**, 108741 (2021)
35. Zhang, Z., Yavas, D., Liu, Q., Wu, D.: Effect of build orientation and raster pattern on the fracture behavior of carbon fiber reinforced polymer composites fabricated by additive manufacturing. Addit. Manuf. **47**, 102204 (2021)
36. Yavas, D.: High-temperature fracture behavior of carbon fiber reinforced PEEK composites fabricated via fused filament fabrication. Compos. B: Eng. **266**, 110987 (2023)